New Concepts in Oxidation Processes

New Concepts in Oxidation Processes

Special Issue Editors

Eric Genty
Ciro Bustillo-Lecompte
Cédric Barroo
Renaud Cousin
Jose Colina-Márquez

MDPI • Basel • Beijing • Wuhan • Barcelona • Belgrade

Special Issue Editors

Eric Genty
Starklab/Terraotherm
France

Ciro Bustillo-Lecompte
School of Occupational and
Public Health, Ryerson
University
Canada

Cédric Barroo
Chemical Physics of Materials
and Catalysis (CPMCT),
Université Libre de Bruxelles
Belgium

Renaud Cousin
Unité de Chimie
Environmentale et Interactions
sur le Vivant (UCEIV EA 4492)
France

Jose Colina-Márquez Department
of Chemical Engineering,
Universidad de Cartagena, Sede
Piedra de Bolíva Colombia

Editorial Office
MDPI
St. Alban-Anlage 66
4052 Basel, Switzerland

This is a reprint of articles from the Special Issue published online in the open access journal *Catalysts* (ISSN 2073-4344) from 2018 to 2019 (available at: https://www.mdpi.com/journal/catalysts/special_issues/oxidation_processes).

For citation purposes, cite each article independently as indicated on the article page online and as indicated below:

LastName, A.A.; LastName, B.B.; LastName, C.C. Article Title. *Journal Name* **Year**, *Article Number*, Page Range.

ISBN 978-3-03928-160-2 (Pbk)
ISBN 978-3-03928-161-9 (PDF)

© 2020 by the authors. Articles in this book are Open Access and distributed under the Creative Commons Attribution (CC BY) license, which allows users to download, copy and build upon published articles, as long as the author and publisher are properly credited, which ensures maximum dissemination and a wider impact of our publications.

The book as a whole is distributed by MDPI under the terms and conditions of the Creative Commons license CC BY-NC-ND.

Contents

About the Special Issue Editors vii

Eric Genty, Ciro Bustillo-Lecompte, Jose Colina-Márquez, Cédric Barroo and Renaud Cousin
Editorial: Special Issue "New Concepts in Oxidation Processes"
Reprinted from: *Catalysts* 2019, 9, 878, doi:10.3390/catal9110878 1

Augusto Arce-Sarria, Fiderman Machuca-Martínez, Ciro Bustillo-Lecompte, Aracely Hernández-Ramírez and José Colina-Márquez
Degradation and Loss of Antibacterial Activity of Commercial Amoxicillin with TiO_2/WO_3-Assisted Solar Photocatalysis
Reprinted from: *Catalysts* 2018, 8, 222, doi:10.3390/catal8060222 3

Déyler Castilla-Caballero, Fiderman Machuca-Martínez, Ciro Bustillo-Lecompte and José Colina-Márquez
Photocatalytic Degradation of Commercial Acetaminophen: Evaluation, Modeling, and Scaling-Up of Photoreactors
Reprinted from: *Catalysts* 2018, 8, 179, doi:10.3390/catal8050179 17

Julien Brunet, Eric Genty, Cédric Barroo, Fabrice Cazier, Christophe Poupin, Stéphane Siffert, Diane Thomas, Guy De Weireld, Thierry Visart de Bocarmé and Renaud Cousin
The CoAlCeO Mixed Oxide: An Alternative to Palladium-Based Catalysts for Total Oxidation of Industrial VOCs
Reprinted from: *Catalysts* 2018, 8, 64, doi:10.3390/catal8020064 32

Niina Koivikko, Tiina Laitinen, Anass Mouammine, Satu Ojala and Riitta L. Keiski
Catalytic Activity Studies of Vanadia/Silica–Titania Catalysts in SVOC Partial Oxidation to Formaldehyde: Focus on the Catalyst Composition
Reprinted from: *Catalysts* 2018, 8, 56, doi:10.3390/catal8020056 52

M. V. Grabchenko, N. N. Mikheeva, G. V. Mamontov, M. A. Salaev, L. F. Liotta and O. V. Vodyankina
Ag/CeO_2 Composites for Catalytic Abatement of CO, Soot and VOCs
Reprinted from: *Catalysts* 2018, 8, 285, doi:10.3390/catal8070285 70

Fudong Liu, Hailiang Wang, Andras Sapi, Hironori Tatsumi, Danylo Zherebetskyy, Hui-Ling Han, Lindsay M. Carl and Gabor A. Somorjai
Molecular Orientations Change Reaction Kinetics and Mechanism: A Review on Catalytic Alcohol Oxidation in Gas Phase and Liquid Phase on Size-Controlled Pt Nanoparticles
Reprinted from: *Catalysts* 2018, 8, 226, doi:10.3390/catal8060226 106

About the Special Issue Editors

Eric Genty obtained his PhD in Chemistry from the Université du Littoral Côte d'Opale (ULCO) in 2014. After a two year postdoctoral stay at the Université Libre de Bruxelles (CPMCT service), he continued at the Ecole Nationale de Chimie de Lille (ENSCL) at Unité de Catalyse et Chimie du Solide (UCCS) for one year in a postdoctoral position in order to study the elemental reaction of CO oxidation over Pt-based catalysts. Following these experiences, Eric took a position as R&D Engineer at Starklab/Terraotherm to develop the depollution aspect of the Terrao heat exchanger. He has co-authored over 20 peer-reviewed scientific papers and one patent.

Ciro Bustillo-Lecompte has a multidisciplinary background in the areas of civil, environmental, and chemical engineering. He completed his Bachelor of Engineering at the University of Cartagena, Colombia, in 2008 and obtained his MASc (2012) and PhD (2016) at Ryerson University, Canada. Ciro is a certified Professional Engineer (PEng), Environmental Professional (EP), a Fraternal Member of the Canadian Institute of Public Health Inspectors (CIPHI), and a 2017–2018 Queen Elizabeth Scholar (QES). He is currently an Associate Member in the Environmental Applied Science and Management Graduate Programs, Program Coordinator at the Real Institute, and a Lecturer in the School of Occupational and Public Health at Ryerson University. He has co-authored over 20 peer-reviewed scientific papers, as well as several conference proceedings, chapters, and books. His research interests include advanced oxidation processes, advanced treatment of water and wastewater, waste minimization, water reuse, water, soil and air quality, energy and resource recovery, and heterogeneous catalysis.

Cédric Barroo obtained his PhD in Chemistry from the Université Libre de Bruxelles in 2014. After a two year postdoctoral stay at Harvard University, he is currently a postdoctoral student at the Université Libre de Bruxelles. His research focuses on the imaging and characterization of catalytic processes using in situ microscopy techniques.

Renaud Cousin (Professor) received his Ph.D. degree in Spectroscopy and Chemistry from Littoral Côte d'Opale University in Dunkirk, France, in 2000, on the topic of "Soot Oxidation". After a postdoctoral position at the University of Strasbourg, sponsored by Daimler, He worked as Assistant Professor at the Littoral Côte d'Opale University, France, from 2003 to 2016. In 2014, he obtained an accreditation to Supervise Research (Habilitation Thesis). In 2016 he was promoted to Full Professor. Currently, his research focuses on the development and characterization of heterogeneous catalysts, for application to the elimination of environmental pollutants (Soot, CO, VOCs, ...). He has co-authored over 70 peer-reviewed scientific papers and one patent.

José Colina-Marquez has been an Associate Professor in the Chemical Engineering Department of the University of Cartagena, since 2010. He obtained his B.Sc. in Chemical Engineering in the University of Atlántico (1996) and his M. Sc. and Ph. D in Chemical Engineering in the University of Valle (2008 and 2010, respectively). Currently, he is leading the Research Group of Modeling and Applications of Advanced Oxidation Processes, that aims to solve water detoxification issues using these technologies. He is also a member of the Editorial Committee of the Revista Ingeniería y Competitividad (University of Valle, 2012) and a member of the Editorial Committee of the Revista Ciencias e Ingeniería (University of Cartagena, 2011). He was awarded with the "Magna cum laude" grade for his PhD studies, granted by the University of Valle (2010), and "Junior Researcher of the year", granted by the Colombian Society of Catalysis (2012).

Editorial

Editorial: Special Issue "New Concepts in Oxidation Processes"

Eric Genty [1,2,*], Ciro Bustillo-Lecompte [3,4], Jose Colina-Márquez [5], Cédric Barroo [2,6,*] and Renaud Cousin [1,*]

1. Unité de Chimie Environnementale et Interactions sur le Vivant, Université du Littoral Côte d'Opale, MREI1—145 Avenue Maurice Schumann, 59140 Dunkerque, France
2. Chemical Physics of Materials and Catalysis, Université Libre de Bruxelles, Faculty of Sciences, Campus Plaine CP 243, 1050 Brussels, Belgium
3. School of Occupational and Public Health, Ryerson University, 350 Victoria Street, Toronto, ON M5B 2K3, Canada; ciro.lecompte@ryerson.ca
4. Graduate Programs in Environmental Applied Science and Management, Ryerson University, 350 Victoria Street, Toronto, ON M5B 2K3, Canada
5. Chemical Engineering Program, Universidad de Cartagena, Av. El Consulado 48-152, Cartagena A.A. 130001, Colombia; jcolinam@unicartagena.edu.co
6. Interdisciplinary Center for Nonlinear Phenomena and Complex Systems (CENOLI), Université Libre de Bruxelles, CP 231, 1050 Brussels, Belgium
* Correspondence: eric.genty@univ-littoral.fr (E.G.); cbarroo@ulb.ac.be (C.B.); renaud.cousin@univ-littoral.fr (R.C.)

Received: 16 October 2019; Accepted: 17 October 2019; Published: 23 October 2019

Oxidation processes, as part of the catalysis field, play a significant role in both industrial chemistry and environmental protection. Without a doubt, the total oxidation reactions of volatile organic compounds (VOCs) and hydrocarbons are critical for environmental pollution prevention and control. Nevertheless, the high incidence of a blend of organic and inorganic compounds (e.g., CO, NOx, SOx, VOC, among others) increases the difficulty of obtaining active, stable, and selective catalytic materials for total oxidation. Another way to eliminate these pollutants is through their selective oxidation to produce highly valuable chemical compounds, such as fuels and alcohols. This approach has also been utilized to yield chemical compounds from biomass. Furthermore, advances in photocatalysis and plasma catalysis permit the intensification of low-energy processes.

The relevance of oxidation processes in the field of environmental catalysis is stimulating interest, as proved by the multiplication of successful Special Issues on this very topic in *Catalysts*:

- *Catalytic Oxidation in Environmental Protection*;
- *New Developments in Heterogeneous Partial and Total Oxidation Catalysis*;
- *Novel Heterogeneous Catalysts for Advanced Oxidation Processes (AOPs)*;
- *Trends in Catalytic Advanced Oxidation Processes*;
- *Photocatalytic Oxidation/Ozonation Processes*;
- *Environmental Catalysis in Advanced Oxidation Processes*;
- *Heterogeneous Catalysis and Advanced Oxidation Processes (AOP) for Environmental Protection (VOCs Oxidation, Air and Water Purification)*;

This Special Issue is focusing on "New Concepts in Oxidation Processes" and aims to cover recent and novel advancements as well as future trends in the field of catalytic oxidation reactions. Topics addressed in this Special Issue include the influence of different parameters on catalytic oxidation at various scales (atomic, laboratory, pilot, or industrial scale), the development of new catalytic materials of environmental or industrial importance, as well as the development of new methods to analyze oxidation processes. A total of six papers were published, covering different aspects of oxidation catalysis. Two papers are focused on photocatalysis. The first one proved that the calcination temperature has a significant effect on the photocatalytic performance for removing amoxicillin, leading to the formation of oxidation byproducts and to the decrease of amoxicillin antibiotic activity [1]. The second paper, combining experiments and theory, emphasizes the degradation of commercial acetaminophen [2]. The use of heterogeneous catalysts is then highlighted in the frame of total oxidation of industrial VOCs using a CoAlCeO mixed-oxides catalyst as an alternative to precious-metal-based materials [3], but also for partial oxidation of sulfur-containing volatile organic compound (SVOC) using vanadia-based catalysts, proving the significant role of the composition of the support in the catalytic behavior [4].

Two review papers complete this Special Issue. The first one summarizes the recent advances and trends on the role of metal–support interactions in Ag/CeO$_2$ composites in their catalytic performance for the total oxidation of CO, soot, and VOCs, and the promising photo- and electro-catalytic applications [5]. The second one consists of a systematic study of catalytic alcohol oxidation on size-controlled platinum nanoparticles in both gas and liquid phases [6] and demonstrates that different molecular orientations in gas and liquid phases lead to very distinct reaction kinetics and mechanisms.

Given these diverse contributions, it is evident that catalytic oxidation processes will continue to flourish. There are still many fundamental questions that remain unanswered, promising a great future for this field. Finally, the Guest Editors would like to sincerely thank all the authors for their valuable contributions.

Conflicts of Interest: The authors declare no conflict of interest.

References

1. Arce-Sarria, A.; Machuca-Martínez, F.; Bustillo-Lecompte, C.; Hernández-Ramírez, A.; Colina-Márquez, J. Degradation and Loss of Antibacterial Activity of Commercial Amoxicillin with TiO$_2$/WO$_3$-Assisted Solar Photocatalysis. *Catalysts* **2018**, *8*, 222. [CrossRef]
2. Castilla-Caballero, D.; Machuca-Martínez, F.; Bustillo-Lecompte, C.; Colina-Márquez, J. Photocatalytic Degradation of Commercial Acetaminophen: Evaluation, Modeling, and Scaling-Up of Photoreactors. *Catalysts* **2018**, *8*, 179. [CrossRef]
3. Brunet, J.; Genty, E.; Barroo, C.; Cazier, F.; Poupin, C.; Siffert, S.; Thomas, D.; De Weireld, G.; Visart de Bocarmé, T.; Cousin, R. The CoAlCeO Mixed Oxide: An Alternative to Palladium-Based Catalysts for Total Oxidation of Industrial VOCs. *Catalysts* **2018**, *8*, 64. [CrossRef]
4. Koivikko, N.; Laitinen, T.; Mouammine, A.; Ojala, S.; Keiski, R.L. Catalytic Activity Studies of Vanadia/Silica–Titania Catalysts in SVOC Partial Oxidation to Formaldehyde: Focus on the Catalyst Composition. *Catalysts* **2018**, *8*, 56. [CrossRef]
5. Grabchenko, M.V.; Mikheeva, N.N.; Mamontov, G.V.; Salaev, M.A.; Liotta, L.F.; Vodyankina, O.V. Ag/CeO$_2$ Composites for Catalytic Abatement of CO, Soot and VOCs. *Catalysts* **2018**, *8*, 285. [CrossRef]
6. Liu, F.; Wang, H.; Sapi, A.; Tatsumi, H.; Zherebetskyy, D.; Han, H.-L.; Carl, L.M.; Somorjai, G.A. Molecular Orientations Change Reaction Kinetics and Mechanism: A Review on Catalytic Alcohol Oxidation in Gas Phase and Liquid Phase on Size-Controlled Pt Nanoparticles. *Catalysts* **2018**, *8*, 226. [CrossRef]

© 2019 by the authors. Licensee MDPI, Basel, Switzerland. This article is an open access article distributed under the terms and conditions of the Creative Commons Attribution (CC BY) license (http://creativecommons.org/licenses/by/4.0/).

Article

Degradation and Loss of Antibacterial Activity of Commercial Amoxicillin with TiO$_2$/WO$_3$-Assisted Solar Photocatalysis

Augusto Arce-Sarria [1], Fiderman Machuca-Martínez [1], Ciro Bustillo-Lecompte [2], Aracely Hernández-Ramírez [3] and José Colina-Márquez [4],*

1. Escuela de Ingeniería Química, Universidad del Valle, Cali A.A. 25360, Colombia; augusto.arce@correounivalle.edu.co (A.A.-S.); fiderman.machuca@correounivalle.edu.co (F.M.-M.)
2. School of Occupational and Public Health, Ryerson University, 350 Victoria Street, Toronto, ON M5B 2K3, Canada; ciro.lecompte@ryerson.ca
3. Facultad de Ciencias Químicas, Universidad de Nuevo León, CP 64570 Monterrey, Nuevo Leon, Mexico; aracely.hernandezrm@uanl.edu.mx
4. Chemical Engineering Program, Universidad de Cartagena, Av. El Consulado 48-152, Cartagena A.A. 130001, Colombia
* Correspondence: jcolinam@unicartagena.edu.co; Tel.: +57-311-788-1188

Received: 30 April 2018; Accepted: 21 May 2018; Published: 23 May 2018

Abstract: In this study, a TiO$_2$ catalyst, modified with tungsten oxide (WO$_3$), was synthesized to reduce its bandgap energy (E$_g$) and to improve its photocatalytic performance. For the catalyst evaluation, the effect of the calcination temperature on the solar photocatalytic degradation was analyzed. The experimental runs were carried out in a CPC (compound parabolic collector) pilot-scale solar reactor, following a multilevel factorial experimental design, which allowed analysis of the effect of the calcination temperature, the initial concentration of amoxicillin, and the catalyst load on the amoxicillin removal. The most favorable calcination temperature for the catalyst performance, concerning the removal of amoxicillin, was 700 °C; because it was the only sample that showed the rutile phase in its crystalline structure. Regarding the loss of the antibiotic activity, the inhibition tests showed that the treated solution of amoxicillin exhibited lower antibacterial activity. The highest amoxicillin removal achieved in these experiments was 64.4% with 100 ppm of amoxicillin concentration, 700 °C of calcination temperature, and 0.1 g L^{-1} of catalyst load. Nonetheless, the modified TiO$_2$/WO$_3$ underperformed compared to the commercial TiO$_2$ P25, due to its low specific surface and the particles sintering during the sol-gel synthesis.

Keywords: sol-gel; bandgap energy; CPC; emergent pollutants; photodegradation

---

1. Introduction

Heterogeneous photocatalysis, based on TiO$_2$, has been widely used for environmental applications such as removal of contaminants and water disinfection due to its oxidative reactions [1–3]. However, TiO$_2$ shows a significant limitation when solar radiation is used for promoting the formation of oxidant species including hydroxyl radicals (•OH) because TiO$_2$ uses only a small fraction of the electromagnetic spectrum corresponding to the UV (Ultraviolet) radiation (wavelengths shorter than 400 nm) [4,5]. To improve the usage of the solar spectrum, several alternatives have been proposed, including catalyst doping, dye-sensitization, and modification with other oxides [6,7].

Ramos-Delgado et al. [8,9] observed the highest photocatalytic activity of TiO$_2$/WO$_3$ materials while using 1% w/w of WO$_3$. Thus, the selection of WO$_3$ as modifying oxide is encouraged by the reduction of the bandgap energy (E$_g$ = 2.8 eV), which has also been reported for TiO$_2$ in a previous work [10].

The reduction of the E_g improves the radiation usage by the photocatalyst since WO_3 can act as an electron-accepting species and reduces the recombination rate of the electron-hole pairs. Regarding the photocatalytic mechanism, the semiconductor TiO_2 is responsible for the electron exchange in the redox reactions and WO_3 can act as a defect of the crystalline structure, inserting an energetic localized state [8,9].

For assessing the photocatalytic activity of the TiO_2/WO_3 material, the oxidation of a commercial antibiotic (amoxicillin) was studied in the presence of solar radiation. Amoxicillin is one of the most consumed antibiotics worldwide and concentrations in the range of 3–87 µg L^{-1} have been reported for hospital effluents [11]. In general, antibiotics have been classified as emergent pollutants due to the potential risks involved with their presence in water bodies and the recent interest in looking for treatment alternatives for their removal. The highest environmental risk of these drugs is the development of waterborne pathogens resistant to the antibiotic activity. Therefore, their natural resistance to biological wastewater treatments has directed the research to novel and more effective technologies for removing these pollutants [12].

Amoxicillin is recognized to be highly refractory and persistent in aquatic ecosystems. Due to the non-selective nature of •OH, several emergent contaminants, including amoxicillin, can be entirely oxidized by advanced oxidation processes (AOPs) as previously reported [13–15]. Photo-Fenton has been reported as an alternative for amoxicillin removal, achieving 52% of total organic carbon (TOC) reduction [16]. Regarding heterogeneous photocatalysis, few applications with TiO_2/WO_3 as a catalyst have been reported. Ramos-Delgado et al. [8] synthesized TiO_2 modified with WO_3 for degrading Malathion, an organophosphorus pesticide. The TOC removal in this work was 78%, comparable with the 47% removal obtained with bare TiO_2. It is important to note that for TiO_2/WO_3, there are no reports of antibiotics removal. However, there are previous studies of TiO_2 doped with Fe and C where 78% of amoxicillin removal was achieved [12,17], evidencing the satisfactory performance of the photocatalysis for eliminating amoxicillin.

This work assessed the photocatalytic activity, not only based on the amoxicillin degradation or the TOC removal but also estimating the loss of the antibacterial activity. It has been found that despite achieving a complete degradation of amoxicillin, even with high TOC removals, the presence of the remaining intermediates can show some antibacterial activity [18]. Regarding bacterial inactivation, this can be more harmful than the presence of the parent antibiotic since waterborne bacteria may develop a more effective resistance to antibiotic activity. Nevertheless, there is no information about the survival or regrowth rates for specific bacteria in such conditions.

Regarding the use of solar radiation as photon source, this is precisely one of the advantages of the reduction of the E_g for the TiO_2/WO_3-based photocatalysis [9,19,20]. It is expected to observe a better performance of the modified TiO_2 in comparison with bare TiO_2 due to a broader absorption of the radiation spectrum of the modified photocatalyst, as mentioned earlier. The experiments of this research were carried out in a pilot-scale CPC photoreactor [21] under the tropical weather conditions of Cali, Colombia, to evaluate the activity of the modified TiO_2 with solar radiation for removing commercial amoxicillin. Moreover, the kinetics was studied by fitting the parameters of a modified Langmuir-Hinshelwood expression with experimental data gathered from the solar photocatalytic tests. The accumulated UV energy was chosen as the independent variable instead of time in this kinetic analysis, because of the variation of the solar irradiation during the experimental runs. This approach allows a consideration of a more accurate manner of a potential scale-up of the photoreactor since the photocatalytic reaction rate depends on the photon absorption as it has been reported in previous studies [22,23].

2. Results

2.1. Effect of the Calcination Temperature on the TiO₂/WO₃ Characterization

The Kubelka-Munk theory was applied to obtain the E_g and hence the absorption wavelength of the material [19–21,24–26]. Figure 1 shows that the lowest reflectance (highest UV absorbance) was observed for the sample calcined at 500 °C. Ramos-Delgado et al. [8] synthesized TiO₂/WO₃ (2% w/w) using the same calcination temperature, obtaining satisfactory results in terms of the particle size and the E_g. Although it was expected to obtain higher reflectance values at higher temperatures, the performance with a calcination temperature of 700 °C shows an intermediate reflectance. As reported in other studies [27,28], this can be related to the ratio of anatase/rutile present in the synthesized material. The calcination temperature can affect the formation of determined crystalline phase and the ratio of these phases [29]. Although it is reported that rutile is the most photoactive phase, it is also the most unstable. The rutile phase appears at temperatures higher than 600 °C; therefore, the lower transmittance observed at 700 °C can be attributed to this phenomenon.

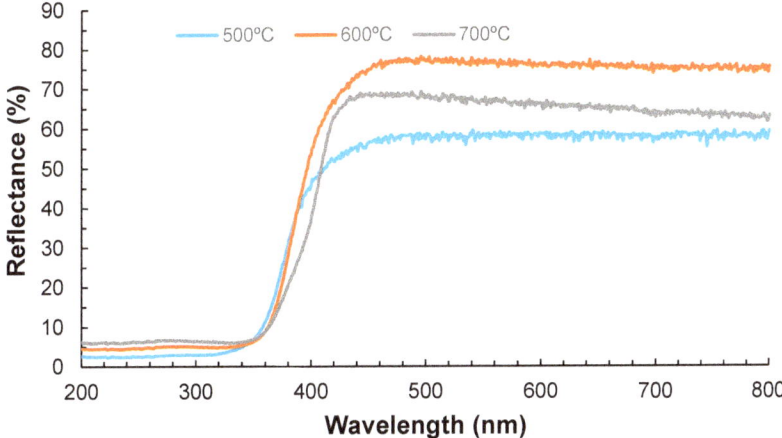

Figure 1. DRS (diffuse reflectance spectroscopy)-UV Vis spectra for 1% TiO₂/WO₃ at different temperatures.

The Kubelka-Munk function (Equation (1)) was used for estimating the E_g based on the reflectance values obtained in Figure 1 for each synthesized material, as follows [30]:

$$F(R_\infty) = \frac{(1-R_\infty)^2}{2R_\infty} \quad (1)$$

where R_∞ corresponds to the ratio between the sample reflectance and a blank reflectance measured in the same equipment. These values are not shown in the manuscript due to the high amount of data obtained from the DRS analysis. The E_g could be calculated with the following equation:

$$[F(R_\infty)h\nu]^{0.5} = C_2(h\nu - E_g) \quad (2)$$

The plot of $[F(R_\infty)h\nu]^{0.5}$ vs. $h\nu$ (Figure 2) allowed to estimate the E_g based on the intercept of the tangent of the obtained curve. For the case of the sample calcined at 700 °C, the obtained value of E_g was 2.84 eV. The E_g results and the maximal wavelength of absorbed radiation for the samples calcined at different temperatures are shown in Table 1.

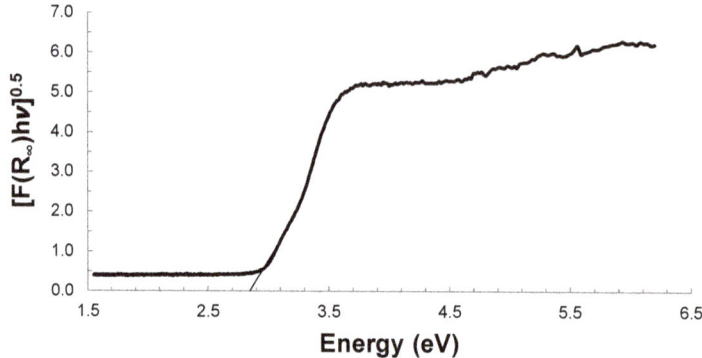

Figure 2. E_g determination for photocatalyst calcined at 700 °C.

Table 1. Bandgap energy and maximum wavelength of radiation absorption.

	Calcination Temperature		
	500 °C	600 °C	700 °C
E_g (eV)	3.12	3.12	2.84
λ (nm)	397	397	436

The reduction of the E_g with respect to bare TiO_2 (3.2 eV) is related to the modification of its crystalline structure due to insertion of the WO_3. The function of this oxide is to add a localized state into the energy gap between the conduction and the valence bands. Furthermore, the insertion of the WO_3 (an electron acceptor species) increases the density of energy holes or vacancies on the TiO_2 surface, and this prevents the electron-hole recombination [8]. As seen in Table 1, the best E_g was obtained at 700 °C, and this result is consistent with the one observed in Figure 1. The E_g values for 500 and 600 °C were the same, but slightly lower than the corresponding one to bare TiO_2. Higher calcination temperatures can promote the formation of the rutile crystalline phase, which has a lower E_g than the anatase phase. However, these changes could not be detected by XRD (X-ray Diffraction) (Figure 3) due to the low concentrations of the WO_3.

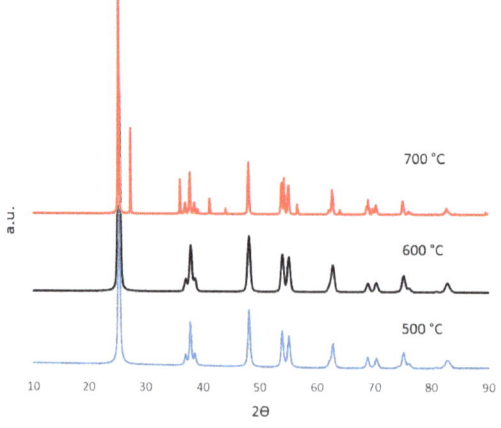

Figure 3. XRD for different synthesized materials.

The difference between the bandgap energies obtained at higher calcination temperatures can be attributed to the characteristic retention of the OH groups by the solids prepared by the sol-gel method [8]. Because of the E_g reduction, compared to bare TiO_2, TiO_2/WO_3 can absorb radiation under 465 nm of wavelength, which means that the material can use part of the visible spectrum of light, as reported in previous works [9,19,20,23,31,32].

Figure 3 shows the XRD patterns obtained with the three different calcination temperatures. There is only one crystalline phase at 500 and 600 °C corresponding to the anatase (tetragonal) structure [8]; whereas, at 700 °C two phases appear, corresponding to a mixture of anatase (JCPDS 98-009-6394) and rutile (JCPDS 98-004-1028) structures [33]. This result is consistent with those shown in Figure 1, where the DRS obtained at 700 °C showed a lower reflectance than the one obtained at 600 °C. This outcome is congruent with the reported literature [8–10] since the rutile phase has a lower E_g than the anatase phase, as mentioned previously. Regarding WO_3, its presence could not be detected by XRD because of its low content in the photocatalyst [10].

The values of the crystal diameter (perpendicular and parallel) and the proportions of the phases were obtained by processing the data with the X'Pert (Malvern Panalytical, Malvern, United Kingdom), GSAS (Edgewall Software, Pittsburgh, PN, USA), and EXPGUI (Edgewall Software, Pittsburgh, PN, USA) software packages, as seen in Table 2.

Table 2. Crystal diameters.

	500 °C	600 °C	700 °C	
	Anatase	Anatase	Anatase	Rutile
$\varnothing_{perp}$ (nm)	59	58	116	820
$\varnothing_{para}$ (nm)	83	37	137	204

It is important to note that the sample calcined at 700 °C exhibited an anatase/rutile ratio: 74/26; nonetheless, the average crystal diameters are much larger than the obtained ones at 500 and 600 °C. The large crystal sizes are the product of the clustering of the WO_3 on the TiO_2 surface as reported in similar studies [8,9]. Although the presence of these clusters can be beneficial for the photocatalytic activity since they can avoid the hole-electron recombination, a larger crystal may affect the performance of the material in photocatalytic reactions negatively, because of the significant decrease of the surface area.

The results in Table 3 are the logical consequence of the behavior observed in Table 2. As the crystal size increases, the surface area decreases as expected. The significant reduction of the surface area for the sample calcined at 700 °C may be related to the formation of WO_3 clusters mentioned previously.

Table 3. Surface area and average pore diameter.

Calcination Temperature (°C)	Surface Area ($m^2\ g^{-1}$)	Average Pore Diameter (nm)
500	66.45	77.53
600	35.93	77.31
700	4.970	122.40

Regarding the pore diameter, the results for the catalysts calcined at the different temperatures show similar diameters for the samples obtained at 500 and 600 °C (~77 nm); however, a much larger diameter (122.40 nm) was exhibited for the 700 °C sample. This last result represents a potential positive effect for the photocatalytic reaction because the mass transport through the catalyst pore will be easier than in smaller pores. The decrease of the surface area, with the subsequent increase of the pore size, can be explained due to the material sintering during the calcination at higher temperatures.

From the obtained results after carrying out physical adsorption tests with nitrogen, it can be said that the solids are considered as mesoporous. This outcome was confirmed by the presence of hysteresis in the adsorption and desorption processes, as seen in Figure 4.

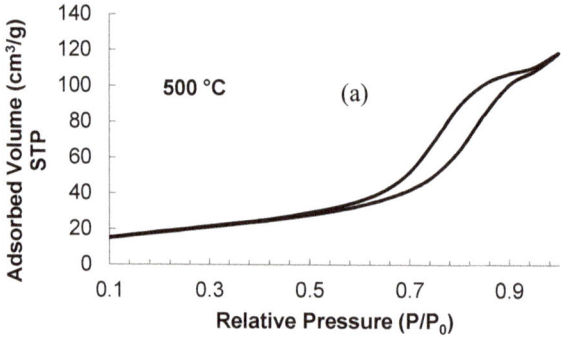

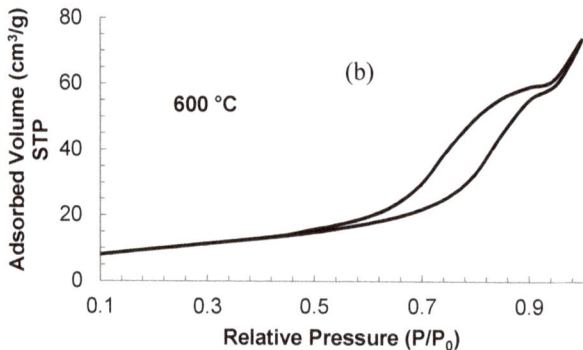

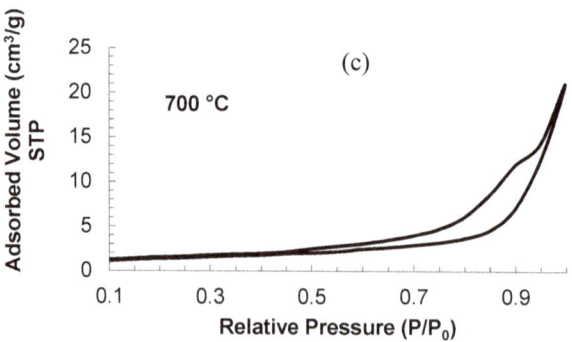

Figure 4. Absorption isotherms for material calcined at different temperatures. (**a**) 500 °C; (**b**) 600 °C; (**c**) 700 °C.

The curves in Figure 4 indicate that the isotherms of the 500 and 600 °C samples are type V; whereas the isotherm for the sample calcined at 700 °C is more similar to a type III [34]. As mentioned above, this is a consequence of the pore diameter of the solid. On the other hand, it can be observed that the adsorbed volume is larger for the sample calcined at 500 °C, which is congruent with the specific surface area estimated by the Brunauer, Emmet, and Teller (BET) method (Table 3).

The thermogravimetric analysis (Figure 5) was carried out to analyze the effect of the temperature on the chemical stability of the material after the programmed heating of the samples without calcining.

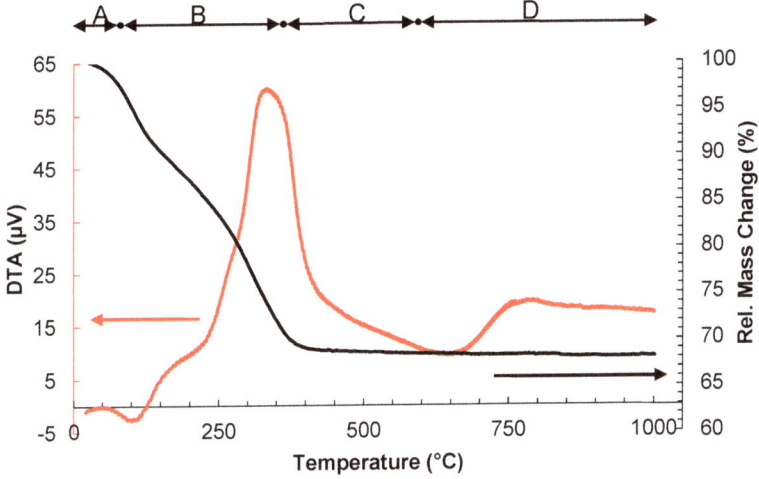

Figure 5. Differential thermogravimetric analysis and thermogram for synthesized material.

From Figure 5, four different regions are well differenced from the differential thermal analysis (DTA): (A) loss of adsorbed water molecules under 150 °C; (B) Elimination of the precursors (sec-butanol, tert-butoxide, and glacial acetic acid) and chemisorbed water from 150 to 400 °C; (C) Formation of TiO_2 crystals from 400 to 600 °C; and (D) Stable weight loss over 600 °C [35]. This outcome supports the results of the XRD analysis, where the peak of the rutile phase appeared at 700 °C as reported in the literature [36,37].

2.2. Degradation and Loss of Antibacterial Activity of Commercial Amoxicillin by TiO_2/WO_3-Assisted Solar Photocatalysis

The results of the experimental design for the amoxicillin degradation are shown in Table 4: The highest amoxicillin degradation was achieved with the sample calcined at 700 °C, an initial amoxicillin concentration of 100 ppm and a catalyst load of 0.10 g L^{-1}.

Table 4. Amoxicillin solar photocatalytic degradation.

		Calcination Temperature, °C					
		500		600		700	
Catalyst load [g L^{-1}]		0.05	0.10	0.05	0.10	0.05	0.10
Amoxicillin concentration [ppm]	100	39.8	28.6	58.6	31.7	45.6	64.4
	200	4.7	16.7	17.6	46.0	12.0	17.0

2.2.1. Effect of the Calcination Temperature

The most relevant fact that can favor the degradation of amoxicillin is that with a calcination temperature above 650 °C the rutile phase appears, and the photocatalytic activity of the synthesized material increases. This fact was evidenced on the XRD of Figure 3, which shows a small peak next to the anatase main peak.

As discussed before, the ratio of anatase/rutile of the sample calcined at 700 °C was found to be 74/26, which is very similar to that reported for the commercial TiO_2 Aeroxide P25 (Evonik, Essen, Germany) [28]. Although the surface area of this sample was the lowest of the three materials tested (due to the TiO_2 sintering at higher temperatures), the larger pore diameter seems to compensate this significant drawback of the catalyst. The sample calcined at 600 °C showed quite good performance as well, which suggests that there must be an optimum of calcination temperature between 600 and 700 °C. Further experiments should be carried out to synthesize a material not only with adequate surface area and particle size but also with a good photoactivity due to the rutile phase presence.

Regarding the E_g, the sample calcined at 700 °C showed the lowest value and its photocatalytic performance was the best of the three samples tested. This result is congruent with the main objective of the TiO_2 modification, which is to reduce the bandgap energy and to improve the photocatalytic activity.

2.2.2. Effect of the Initial Amoxicillin Concentration

The higher initial concentrations of the substrate in any photocatalytic reaction negatively affect the catalyst performance, as has been reported in several works [16,22,28]. In this study, the same behavior was observed as well.

The higher concentrations of amoxicillin are detrimental to the photocatalytic degradation because of the reduction of the available active sites of the catalyst after the adsorption of the amoxicillin and other compounds to the TiO_2/WO_3 surface. In the case of higher concentrations, the adsorbed molecules can inhibit the •OH radicals' generation and the degradation rate decreases as a logical consequence.

2.2.3. Effect of the Catalyst Load

The catalyst load can have a positive effect on the photocatalytic degradation; that means that the performance will increase with an increase of the catalyst load as can be observed from the results shown in Table 4.

Nonetheless, the presence of a maximum has been reported in previous studies [22,38–40], which is around 0.35 g L^{-1} for a CPC reactor of the same characteristics used for this research but with TiO_2 P25 as the catalyst. The existence of this maximum is because of the "clouding" effect in the reactor when higher catalyst loads are used in photocatalytic reactions. This phenomenon occurs when an excessive number of particles suspended in the reactor does not allow the photon to pass through the bulk liquid, and therefore, it avoids the generation of the electron-hole pairs necessary for the •OH formation.

The highest catalyst load used in these experiments (0.10 g L^{-1}) is still lower than the maximum mentioned above; consequently, it is expected that better degradations are achieved with this value. From Table 4, it can be observed that only the samples calcined at 700 °C show this behavior.

As mentioned above, the sample of 700 °C exhibited the largest crystal size and accordingly, the largest cluster size. With these characteristics, they exhibit less scattering and photon absorption due to their larger size and thus, higher catalyst loads are required to generate the same amount of •OH radicals than the solids with smaller sizes (samples calcined at 500 and 600 °C) [38–40].

Although the effect of the pH was not considered in this study since the experiments were carried out at the natural pH of the solution (6.8–7.0), it is reported that the particle size is affected by the pH [5]. When the solution pH is close to the zero-charge point (pH_{zpc}) of the solid, this tends to form large clusters with the consequences mentioned above. For the commercial TiO_2, the reported pH_{zpc} is around 6.5 [41]; therefore, it is probable that the value for the synthesized material in this study is similar and large clusters are formed.

2.2.4. Kinetic Analysis of the Amoxicillin Photocatalytic Degradation

The TiO$_3$/WO$_3$ calcined at 700 °C was chosen for the comparative kinetic study of the solar photocatalytic degradation of commercial amoxicillin. Figure 6 shows a comparison between the performance of this catalyst and the Aeroxide P25; both tested at the following reaction conditions: 100 ppm of the initial concentration of amoxicillin, 0.1 g L^{-1} for catalyst load and 510,000 J m^{-2} of UVA (Ultraviolet A) accumulated radiation.

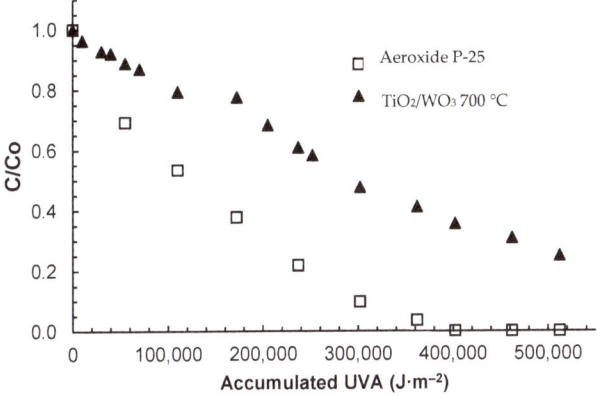

Figure 6. Photocatalytic degradation of amoxicillin with TiO$_2$/WO$_3$ and Aeroxide P25.

The accumulated UVA energy was set as the independent variable instead of time because of the variability of the solar irradiation. Consequently, the kinetic law will be expressed in terms of the accumulated radiation that arrives at the solar reactor.

From Figure 6, it is evident that the P25 exhibited better performance than the TiO$_2$/WO$_3$. While the anatase/rutile ratio is very similar for both catalysts, the difference between the crystal sizes and the surface area are significant (50 m^2 g^{-1} for P25 vs. 4.97 m^2 g^{-1} for TiO$_2$/WO$_3$). The sintering of the TiO$_2$/WO$_3$ particles at high calcination temperatures may be the primary cause of the formation of large clusters, as discussed previously.

On the other hand, the E$_g$ of the TiO$_2$/WO$_3$ is lower than the one of the P25; therefore, it was expected to have a higher photocatalytic degradation rate for the modified material since it could absorb photons of the visible part of the solar radiation spectrum.

Nonetheless, a lower E$_g$ was not enough to improve the material performance over the P25 regarding photocatalytic applications. This result suggests that the modification of a semiconductor should not only be focused on decreasing the bandgap energy, but also on improving other properties that can affect the photocatalytic performance such as the surface area, pore diameter or the presence of photoactive crystalline phases significantly.

For analyzing the kinetics of the photocatalytic degradation, a modified Langmuir-Hinshelwood (L-H) was used and after fitting the model parameters. The obtained results are shown as follows in Table 5:

Table 5. Apparent kinetic and adsorption parameters of the modified L-H model.

Catalyst	K_{ads} (ppm^{-1})	k_{app} (ppm m^2 J^{-1})
TiO$_2$/WO$_3$	0.0632	1.88×10^{-4}
Aeroxide P25	0.0087	1.14×10^{-3}

Results shown in Table 5 can be related to the difference between the specific surface area of the catalysts. Where the TiO_2/WO_3 showed a higher value of the adsorption constant than the P25, the latter exhibited a higher apparent kinetic constant. The global reaction rate is limited by the kinetic for the case of the TiO_2/WO_3 catalyst because of the surface can be covered easily by the substrates (that means a zero-order rate law). Whereas, for the P25, the behavior of the degradation kinetics fits a pseudo-first order law trend, where the adsorption can limit the rate or the kinetics indistinctly [41].

2.2.5. Loss of Antibiotic Power

Figure 7 shows a typical antibiogram, corresponding to the sample obtained at the most favorable conditions for the photocatalytic degradation of amoxicillin. This result suggests that the photo-oxidized amoxicillin could be transformed into a compound with less antibiotic activity.

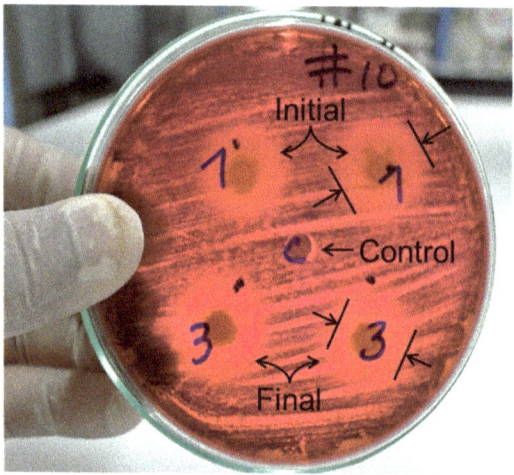

Figure 7. Antibiogram; (1) and (3) correspond to initial, and final simple for a treatment, (C) is a control.

It has been reported that the photocatalytic degradation of several pharmaceuticals can lead to more toxic byproducts [11,13,16,17]. Concerning antibiotics, the disappearance of the parent compound does not guarantee that the byproducts do not present antibacterial activity, despite the non-selective nature of the photocatalytic oxidation.

In Figure 7, the inhibition halo for the sample after photocatalytic treatment (3) is smaller than that observed for the sample before the treatment (1). From this observation, it can be proposed that the photocatalytic degradation yielded a lower number of antibiotic byproducts. This outcome is congruent with previous works that used TiO_2-assisted photocatalysis for removing amoxicillin and other antibiotics [11–13,16,17]. This result can be used in future works that study the coupling of heterogeneous solar photocatalysis to a biological system to polish the wastewater treatment.

3. Materials and Methods

3.1. Preparation of TiO_2/WO_3 Catalysts

For synthesizing TiO_2/WO_3, ammonium p-tungstate (ApT) (Sigma-Aldrich, Cali, Colombia), tetrabutyl orthotitanate (TBT) (Fluka, Bogota, Colombia), sec-butanol (SB) (Thermo Fisher Scientific, Bogota, Colombia), and glacial acetic acid (GAA) (Sigma-Aldrich, Cali, Colombia) were used. For dilutions, ultrapure water (MilliQ) with an 18 $M\Omega$ cm^{-1} resistivity (Merck, Cali, Colombia) was used.

The catalyst was prepared by using the sol-gel process, as suggested in the previous studies of Ramos-Delgado et al. [8,9]. For obtaining 10 g of catalyst, 29.1 mL of TBT in 100 mL of SB were mixed. The pH was adjusted to 3.5 GAA units. The pre-hydrolytic treatment was carried out with a mixture of 56.3 mL of SB and 1.15 mL of H$_2$O. Then, the ApT solution (89 mg ApT in 7.5 mL of water) was slowly added dropwise. Finally, the solution was aged for two days; the resulting powder was finely macerated in an agate mortar before calcination. Moreover, three temperature calcination of the material 500, 600 and 700 °C were evaluated.

3.2. Photocatalyst Characterization

A UV-Vis spectrophotometer with diffuse reflectance analyzed the optical absorption of the catalyst (Thermo Fisher Scientific Evolution 300 with integrating sphere, Bogota, Colombia) and used a sample of Spectralon as standard blank. The XRD spectra were obtained with a Siemens D500 XRD equipment (Bogota, Colombia). The crystal size and the crystalline phases present in each of the catalysts were determined by using the Gsas© and Xpert© software packages; while the surface and pore diameter areas and adsorption isotherms were determined with a Quantachrome Autosorb Automated Gas Sorption equipment. Simultaneously, a thermogravimetry analysis (DTA/TG) was performed on a Simultaneous Thermal Analysis (STA) equipment PT1600 TG-DSC/DTA (LINSEIS, Monterrey, Mexico) to determine the behavior of the material in terms of thermal decomposition and phase changes.

3.3. Evaluation of the Photocatalytic Performance

A multilevel factorial experiment design was used for evaluating the following factors: the concentration of catalyst (0.10 g L^{-1} and 0.05 g L^{-1}), the concentration of amoxicillin (100 and 200 ppm), and the calcination temperature (500, 600 and 700 °C). The selected response variable was the degradation of amoxicillin [26,42]. The photocatalytic reaction was carried out in a CPC reactor with 25 L of total volume (Figure 8). The UV accumulated radiation was measured with a UV-radiometer Delta OHM HD210.2 (Bogota, Colombia). For each test, the accumulated energy was fixed at 550,000 J m^{-2}. The commercial amoxicillin for the experimental runs was used as received from 500 mg-capsules (Genfar, Bogota, Colombia). For following the amoxicillin concentration, UPLC analyses were performed using an ACQUITY UPLC BEH® C18 1.7 μm column H-Class (Waters, Bogota, Colombia), with a retention time of 1.36 min. The experimental procedure for carrying out the solar photocatalytic tests was previously reported by Colina-Márquez et al. [41].

Figure 8. Solar pilot-scale CPC reactor (Photocatalysis lab, Universidad del Valle, Cali, Colombia).

3.4. Kinetic Analysis

The Langmuir-Hinshelwood expression was modified to consider the change of the amoxicillin concentrations with respect to the accumulated UVA radiation, as follows:

$$\frac{dC}{dQ_{UV}} = \frac{k_{app}K_{Ads}C}{1+K_{Ads}C} \qquad (3)$$

where C is the amoxicillin concentration in ppm, K_{Ads} is the adsorption constant in ppm^{-1}, Q_{UV} is the accumulated UVA radiation in J m^{-2} and k_{app} is the apparent kinetic constant in ppm·m^2 J^{-1}. For estimating the model parameters, Equation (3) is transformed into its linear form for fitting its parameters from the experimental data by using linear regression:

$$\frac{dQ_{UV}}{dC} = \frac{1}{k_{app}K_{Ads}} \cdot \frac{1}{C} + \frac{1}{k_{app}} \qquad (4)$$

The reciprocal of the term dC/dQ_{UV}, which is analog to dC/dt, was plotted versus the reciprocal of the concentration $(1/C)$. By using the minimum squares analysis, the model parameters, K_{Ads} and k_{app}, were estimated from the slope and the y-intercept of the curve fitted to the experimental results.

3.5. Determination of Inhibition Halo

An antibiogram was performed for determining the potential loss of antibiotic power of the amoxicillin [43]. The reactants used for this test were: eosin Methylene Blue agar (Lot. VM290047 124, Merck, Cali, Colombia), nutrient agar (Lot. 16761, Bogota, Colombia), BactoTM Peptone (Lot. 3063372, Becton Dickinson, Bogota, Colombia) and E. Coli. (WG5) as the model bacterium. The tests were carried out in Petri's boxes of 9 cm and an incubator (WTB Binder, Tuttlingen, Germany) with a McFarland 0.5 standard for determining the concentration of colonies forming units (CFU) of present bacteria.

4. Conclusions

The calcination temperature had a significant effect on the photocatalytic performance for removing amoxicillin. The material synthesized at 700 °C was the only one that exhibited the presence of the rutile crystalline phase in its structure. Nonetheless, it underperformed compared with the commercial standard, Aeroxide P25, which could remove all the amoxicillin with less UVA accumulated radiation than the required one by the TiO_2/WO_3 catalyst. Regarding the loss of antibacterial activity, the inhibition test showed that TiO_2/WO_3-assisted photocatalytic degradation yields oxidation byproducts with less antibiotic activity than the original amoxicillin.

Author Contributions: A.A.-S. carried out the synthesis and the evaluation experiments with solar photocatalysis; F.M.-M. gave the support of his lab for the experimental tests; C.B.-L. helped to structure and write this paper; A.H.-R. provided the support for characterizing the catalysts, and J.C.-M. analyzed and processed the photocatalytic degradation results.

Acknowledgments: The financial and logistic support of Universidad del Valle, Universidad de Cartagena, Universidad Autónoma de Nuevo León, and Ryerson University for their financial and logistic support. A.A. would like to thank Colciencias for his Ph.D. scholarship and J.C.-M. want to thank the Research Office at Universidad de Cartagena for Supporting Recognized Research Groups.

Conflicts of Interest: The authors declare no conflicts of interest.

References

1. Fujishima, A.; Zhang, X.; Tryk, D. Heterogeneous photocatalysis: From water photolysis to applications in environmental cleanup. *Int. J. Hydrogen Energy* **2007**, *32*, 2664–2672. [CrossRef]
2. Alarcon, D.C.; Maldonado, M.I.; Malato, S.; Gernjak, W. Photocatalytic decontamination and disinfection of water with solar collectors. *Catal. Today* **2007**, *122*, 137–149.

3. Domènech, X.; Jardim, W.F.; Litter, M.I. Procesos avanzados de oxidación para la eliminación de contaminantes. In *Eliminación de contaminantes por fotocatálisis heterogénea*; Blesa, M.A., Ed.; CYTED: La Plata, Argentina, 2001; pp. 3–26.
4. Klavarioti, M.; Mantzavinos, D.; Kassinos, D. Removal of residual pharmaceuticals from aqueous systems by advanced oxidation processes. *Environ. Int.* **2009**, *35*, 402–417. [CrossRef] [PubMed]
5. Malato, S.; Fernández-Ibáñez, P.; Maldonado, M.I.; Blanco, J.; Gernjak, W. Decontamination and disinfection of water by solar photocatalysis: Recent overview and trends. *Catal. Today* **2009**, *147*, 1–59. [CrossRef]
6. Zaleska, A. Doped-TiO_2: A Review. *Recent Patents Eng.* **2008**, *2*, 157–164. [CrossRef]
7. Shen, S.H.; Wu, T.Y.; Juan, J.C.; Teh, C.Y. Recent developments of metal oxide semiconductors as photocatalysts in advanced oxidation processes (AOPs) for treatment of dye waste-water. *J. Chem. Technol. Biotechnol.* **2011**, *86*, 1130–1158.
8. Ramos-Delgado, N.A.; Hinojosa-Reyes, L.; Guzman-Mar, J.L.; Gracia-Pinilla, M.A.; Hernández-Ramírez, A. Synthesis by sol–gel of WO_3/TiO_2 for solar photocatalytic degradation of malathion pesticide. *Catal. Today* **2013**, *209*, 35–40. [CrossRef]
9. Ramos-Delgado, N.A.; Gracia-Pinilla, M.A.; Maya-Treviño, L.; Hinojosa-Reyes, L.; Guzman-Mar, J.L.; Hernández-Ramírez, A. Solar photocatalytic activity of TiO_2 modified with WO_3 on the degradation of an organophosphorus pesticide. *J. Hazard. Mater.* **2013**, *263*, 36–44. [CrossRef] [PubMed]
10. Yang, J.; Zhang, X.; Liu, H.; Wang, C.; Liu, S.; Sun, P.; Wang, L.; Liu, Y. Heterostructured TiO_2/WO_3 porous microspheres: Preparation, characterization and photocatalytic properties. *Catal. Today* **2013**, *201*, 195–202. [CrossRef]
11. Elmolla, E.S.; Chaudhuri, M. Photocatalytic degradation of amoxicillin, ampicillin and cloxacillin antibiotics in aqueous solution using UV/TiO_2 and $UV/H_2O_2/TiO_2$ photocatalysis. *Desalination* **2010**, *252*, 46–52. [CrossRef]
12. Das, R.; Sarkar, S.; Chakraborty, S.; Choi, H.; Bhattacharjee, C. Remediation of Antiseptic Components in Wastewater by Photocatalysis Using TiO_2 Nanoparticles. *Ind. Eng. Chem. Res.* **2014**, *53*, 3012–3020. [CrossRef]
13. Fatta-Kassinos, D.; Vasquez, M.I.; Kümmerer, K. Transformation products of pharmaceuticals in surface waters and wastewater formed during photolysis and advanced oxidation processes—Degradation, elucidation of byproducts and assessment of their biological potency. *Chemosphere* **2011**, *85*, 693–709. [CrossRef] [PubMed]
14. Basile, T.; Petrella, A.; Petrella, M.; Boghetich, G.; Petruzzelli, V.; Colasuonno, S.; Petruzzelli, D. Review of Endocrine-Disrupting-Compound Removal Technologies in Water and Wastewater Treatment Plants: An EU Perspective. *Ind. Eng. Chem. Res.* **2011**, *50*, 8389–8401. [CrossRef]
15. Klauson, D.; Babkina, J.; Stepanova, K.; Krichevskaya, M.; Preis, S. Aqueous photocatalytic oxidation of amoxicillin. *Catal. Today* **2010**, *151*, 39–45. [CrossRef]
16. Chaudhuri, M.; Wahap, M.Z.B.A.; Affam, A.C. Treatment of aqueous solution of antibiotics amoxicillin and cloxacillin by modified photo-Fenton process. *Desalin. Water Treat.* **2013**, *51*, 7255–7268. [CrossRef]
17. Dimitrakopoulou, D.; Rethemiotaki, I.; Frontistis, Z.; Xekoukoulotakis, N.P.; Venieri, D.; Mantzavinos, D. Degradation, mineralization and antibiotic inactivation of amoxicillin by $UV-A/TiO_2$ photocatalysis. *J. Environ. Manag.* **2012**, *98*, 168–174. [CrossRef] [PubMed]
18. Frontistis, Z.; Antonopoulou, M.; Venieri, D.; Konstantinou, I.; Mantzavinos, D. Boron-doped diamond oxidation of amoxicillin pharmaceutical formulation: Statistical evaluation of operating parameters, reaction pathways and antibacterial activity. *J. Environ. Manag.* **2017**, *195*, 100–109. [CrossRef] [PubMed]
19. Chen, X. Increasing Solar Absorption for Photocatalysis with Black Hydrogenated Titanium Dioxide Nanocrystals. *Science* **2013**, *331*, 746–750. [CrossRef] [PubMed]
20. Li, X.Z.; Li, F.B.; Yang, C.L.; Ge, W.K. Photocatalytic activity of WO_x-TiO_2 under visible light irradiation. *J. Photochem. Photobiol. A Chem.* **2001**, *141*, 209–217. [CrossRef]
21. Mccullagh, C.; Skillen, N.; Adams, M.; Robertson, P.K.J. Photocatalytic reactors for environmental remediation: A review. *J. Chem. Technol. Biotechnol.* **2011**, *86*, 1002–1017. [CrossRef]
22. Colina-Márquez, J.; Díaz, D.; Rendón, A.; López-Vásquez, A.; Machuca-Martínez, F. Photocatalytic treatment of a dye polluted industrial effluent with a solar pilot-scale CPC reactor. *J. Adv. Oxid. Technol.* **2009**, *12*, 93–99. [CrossRef]
23. Xiaobo, C. Titanium Dioxide Nanomaterials and Their Energy Applications. *Chin. J. Catal.* **2009**, *30*, 839–851.

24. Liu, K.; Hsueh, Y.; Su, C.; Perng, T. Photoelectrochemical application of mesoporous TiO_2/WO_3 nanohoneycomb prepared by sol e gel method. *Int. J. Hydrogen Energy* **2013**, *38*, 7750–7755. [CrossRef]
25. Piszcz, M.; Tryba, B.; Grzmil, B.; Morawski, A.W. Photocatalytic Removal of Phenol Under UV Irradiation on WO_x–TiO_2 Prepared by Sol – Gel Method. *Catal. Lett.* **2009**, *128*, 190–196. [CrossRef]
26. Weng, X.; Chen, Z.; Chen, Z.; Megharaj, M. Colloids and Surfaces A: Physicochemical and Engineering Aspects Clay supported bimetallic Fe/Ni nanoparticles used for reductive degradation of amoxicillin in aqueous solution: Characterization and kinetics. *Colloids Surf. A Physicochem. Eng. Asp.* **2014**, *443*, 404–409. [CrossRef]
27. Haider, A.J.; Anbari, R.H.A.; Kadhim, G.R.; Salame, C.T. Exploring potential Environmental applications of TiO_2 Nanoparticles. *Energy Procedia* **2017**, *119*, 332–345. [CrossRef]
28. Herrera-Barros, A.; Reyes, A.; Colina-Marquez, J. Evaluation of the photocatalytic activity of iron oxide nanoparticles functionalized with titanium dioxide. *J. Phys. Conf. Ser.* **2016**, *687*, 012034. [CrossRef]
29. Yu, J.-G.; Yu, H.-G.; Cheng, B.; Zhao, X.-J.; Yu, J.C.; Ho, W.-K. The Effect of Calcination Temperature on the Surface Microstructure and Photocatalytic Activity of TiO_2 Thin Films Prepared by Liquid Phase Deposition. *J. Phys. Chem. B* **2003**, *107*, 13871–13879. [CrossRef]
30. Escobedo-Morales, A.; Sánchez-Mora, E.; Pal, U. Use of diffuse reflectance spectroscopy for optical characterization of un-supported nanostructures. *Rev. Mex. Fís.* **2007**, *53*, 18–22.
31. Beranek, R.; Kisch, H. Tuning the optical and photoelectrochemical properties of surface-modified TiO_2. *Photochem. Photobiol. Sci.* **2008**, *7*, 40–48. [CrossRef] [PubMed]
32. Patsoura, A.; Kondarides, D.I.; Verykios, X.E. Enhancement of photoinduced hydrogen production from irradiated Pt/TiO_2 suspensions with simultaneous degradation of azo-dyes. *Appl. Catal. B Environ.* **2006**, *64*, 171–179. [CrossRef]
33. Bezrodna, T.; Gavrilko, T.; Puchkovska, G.; Shimanovska, V.; Baran, J. Spectroscopic study of TiO_2 (rutile) –benzophenone heterogeneous systems. *J. Mol. Struct.* **2002**, *614*, 315–324. [CrossRef]
34. Enríquez, J.M.H.; Serrano, L.A.G.; Soares, B.H.Z.; Alamilla, R.G.; Resendiz, B.B.Z.; Del Sánchez, T.; Hernández, , A.C. Síntesis y Caracterización de Nanopartículas de N-TiO_2 – Anatasa. *Superf. y Vacío* **2008**, *21*, 1–5.
35. Rungjaroentawon, N.; Onsuratoom, S.; Chavadej, S. Hydrogen production from water splitting under visible light irradiation using sensitized mesoporous-assembled TiO_2-SiO_2 mixed oxide photocatalysts. *Int. J. Hydrogen Energy* **2012**, *37*, 11061–11071. [CrossRef]
36. Yang, H.; Shi, R.; Zhang, K.; Hu, Y.; Tang, A.; Li, X. Synthesis of WO_3/TiO_2 nanocomposites via sol–gel method. *J. Alloys Compd.* **2005**, *398*, 200–202. [CrossRef]
37. Djaoued, Y.; Badilescu, S.; Ashrit, P.V.; Bersani, D.; Lottici, P.; Robichaud, J. Study of Anatase to Rutile Phase Transition in Nanocrystalline Titania Films. *J. Sol-Gel Sci. Technol.* **2002**, *24*, 255–264. [CrossRef]
38. Colina-Márquez, J.; Machuca, F.; Puma, G.L. Radiation Absorption and Optimization of Solar Photocatalytic Reactors for Environmental Applications. *Environ. Sci. Technol.* **2010**, *44*, 5112–5120. [CrossRef] [PubMed]
39. Mueses, M.A.; Machuca-Martinez, F.; Hernández-Ramirez, A.; Puma, G.L. Effective radiation field model to scattering—Absorption applied in heterogeneous photocatalytic reactors. *Chem. Eng. J.* **2015**, *279*, 442–451. [CrossRef]
40. Nasirian, M.; Lin, Y.P.; Bustillo-Lecompte, C.F.; Mehrvar, M. Enhancement of photocatalytic activity of titanium dioxide using non-metal doping methods under visible light: A review. *Int. J. Environ. Sci. Technol.* **2017**, 1–24. [CrossRef]
41. Colina-Márquez, J.; Machuca-Martínez, F.; Puma, G.L. Photocatalytic mineralization of commercial herbicides in a pilot-scale solar CPC reactor: Photoreactor modeling and reaction kinetics constants independent of radiation field. *Environ. Sci. Technol.* **2009**, *43*, 8953–8960. [CrossRef] [PubMed]
42. Xu, H.; Cooper, W.J.; Jung, J.; Song, W. Photosensitized degradation of amoxicillin in natural organic matter isolate solutions. *Water Res.* **2011**, *45*, 632–638. [CrossRef] [PubMed]
43. Cantón, R.; Gómez-Lus, M.L.; Rodríguez-Avial, C.; Martinez, L.M.; Vila, J. *Procedimientos en Microbiología Clínica*; SIEMC: Madrid, Spain, 2000; pp. 10–20.

© 2018 by the authors. Licensee MDPI, Basel, Switzerland. This article is an open access article distributed under the terms and conditions of the Creative Commons Attribution (CC BY) license (http://creativecommons.org/licenses/by/4.0/).

Article

Photocatalytic Degradation of Commercial Acetaminophen: Evaluation, Modeling, and Scaling-Up of Photoreactors

Déyler Castilla-Caballero [1], Fiderman Machuca-Martínez [1], Ciro Bustillo-Lecompte [2] and José Colina-Márquez [3,*]

1. Escuela de Ingeniería Química, Universidad del Valle, Cali A.A. 25360, Colombia; deyler.castilla@correounivalle.edu.co (D.C.-C.); fiderman.machuca@correounivalle.edu.co (F.M.-M.)
2. School of Occupational and Public Health, Ryerson University, 350 Victoria Street, Toronto, ON M5B 2K3, Canada; ciro.lecompte@ryerson.ca
3. Chemical Engineering Program, Universidad de Cartagena, Av. El Consulado 48-152, Cartagena A.A. 130001, Colombia
* Correspondence: jcolinam@unicartagena.edu.co; Tel.: +57-311-788-1188

Received: 12 April 2018; Accepted: 25 April 2018; Published: 28 April 2018

Abstract: In this work, the performance of a pilot-scale solar CPC reactor was evaluated for the degradation of commercial acetaminophen, using TiO_2 P25 as a catalyst. The statistical Taguchi's method was used to estimate the combination of initial pH and catalyst load while tackling the variability of the solar radiation intensity under tropical weather conditions through the estimation of the signal-to-noise ratios (S/N) of the controllable variables. Moreover, a kinetic law that included the explicit dependence on the local volumetric rate of photon absorption (LVRPA) was used. The radiant field was estimated by joining the Six Flux Model (SFM) with a solar emission model based on clarity index (K_C), whereas the mass balance was coupled to the hydrodynamic equations, corresponding to the turbulent regime. For scaling-up purposes, the ratio of the total area-to-total-pollutant volume (A_T/V_T) was varied for observing the effect of this parameter on the overall plant performance. The Taguchi's experimental design results showed that the best combination of initial pH and catalyst load was 9 and 0.6 g L^{-1}, respectively. Also, full-scale plants would require far fewer ratios of A_T/V_T than for pilot or intermediate-scale ones. This information may be beneficial for reducing assembling costs of photocatalytic reactors scaling-up.

Keywords: photoreactor; modeling; Taguchi; scaling-up; TiO_2; acetaminophen

1. Introduction

Water pollution is a serious threat that has captured the attention of governments and scientific communities worldwide. Chemicals such as pesticides, fertilizers, pharmaceuticals, steroids, disinfectants, preservatives, additives, personal care products, and heavy metals are frequently found in water and wastewaters, and because of this, they are commonly known as emerging contaminants, and some of them are considered persistent organic pollutants (POPs) [1]. Unfortunately, the most common water and wastewater treatment plants are unable to destroy these chemicals due to their recalcitrant nature and toxicity [2].

Among the various adverse effects emerging contaminants and particularly POPs pose to human life and the environment, it is the occurrence of many types of cancer, birth defects, and other developmental disorders due to endocrine disruptors, as well as various diseases such as high blood pressure, renal disorders, joint pain, and malfunctioning of the nervous system, which is strongly related to the intake of water polluted with POPs [3–5]. Acetaminophen was chosen as the model pollutant because it is a massively consumed drug worldwide. The excretion of this drug (in wastewater from hospitals and private households) and disposal of unused medicine have caused acetaminophen to appear in surface waterbodies. A concentration range between 4.6 and 52 µg L^{-1} has been reported in previous studies carried out in several countries of Europe and America [6]. Nonetheless, it is expected that higher concentrations can be found in hospitals wastewaters although there are no reports of these concentrations in literature. Furthermore, water contaminated with pharmaceuticals such as acetaminophen can cause hepatic damage in humans [7,8] and alter the equilibrium of aquatic ecosystems due to its toxicity [9,10]. As a result, there is a need to develop efficient technologies to remove emerging contaminants from industrial wastewater and water bodies.

Heterogeneous photocatalysis has proven to be an effective method for eliminating many emerging pollutants [11–14]. As an advanced oxidation process, heterogeneous photocatalysis uses sunlight as the promotor of the redox reactions responsible for removing the contaminants. The general mechanism of photocatalysis for degrading organic pollutants has been reported in several papers [15,16] and involves the generation of electron-hole pairs, which in the presence of electron acceptors such as atmospheric oxygen, leads to the formation of potent oxidant agents that can destroy and mineralize organic matter.

Although photocatalysis was discovered around three decades ago, the worldwide applications at full-scale are scarce or almost null. The estimation of the total footprint required for operating photocatalytic reactors, as the land occupation is a matter of study from the engineering point of view. To the best of the authors' knowledge, no studies have been reported on the scaling of solar photocatalytic reactors that consider pilot, middle and full-scale schemes, regarding the total area required for its operation. Besides, most of the existing photocatalytic reactors have been designed following empirical methods rather than strict mathematical modeling. This approach can be attributed to the complexity of the modeling and simulation of simultaneous phenomena that take place during its operation, i.e., photonics, hydrodynamics, and kinetics.

Regarding the photon absorption by the catalyst, the estimation of the radiant field can be a challenging task when the catalytic particles are suspended in the reactor due to the scattering that takes place once the photons enter the reactor. This difficulty may be even higher when the solar radiation acts as the photon's source, because of its variability and form of propagation (direct or diffuse). The most common approaches for estimating the radiant field are the Discrete Ordinates Method (DOM) [17,18], the Monte Carlo (MC) simulations [19–21], and the Six-Flux Model (SFM) [21]. The first two methods require far more computational effort than the SFM, which is based mainly on the assumption that when photons collide with catalytic particles, scattering occurs in the six directions of the Cartesian system. Despite the simplicity of the SFM, which is composed only of algebraic equations, several reports show satisfactory fitting to experimental data [21].

Concerning photocatalytic kinetics, the Steady-State Approximation (SSA) is usually applied for the total concentration of holes and hydroxyl radicals (OH$^\bullet$) according to the general mechanism of photocatalysis with TiO$_2$ [22]. This strategy, which was first reported by Turchi and Ollis [15] and Alfano et al. [16], has led to the writing of several critical photo-kinetic laws, such as the modified Langmuir-Hinshelwood equation.

For estimating the effect of the radiation field in the kinetic law, many authors have put explicitly the radiation intensity or the local volumetric rate of photon absorption (LVRPA) in the kinetic equation [23,24], which can allow finding kinetic parameters independent on the reactors' geometry. However, to date, there has not been reported a strategy for finding kinetic parameters that can be suitable for describing the photocatalytic treatment of pollutants independently of the natural variability and propagation form of the solar radiation.

In the present work, a strategy that combines simplicity and accuracy for modeling and simulating solar CPC photocatalytic reactors was used, using acetaminophen as a target molecule. A Langmuir-Hinshelwood-like equation derived from the SSA in the kinetic model was used, which has shown to be adequate for describing many experimental data in the photocatalytic abatement of organic pollutants. Additionally, we applied the Six Flux Model and a solar emission model based on the clarity index (K_C) [25,26] for estimating the LVRPA and coupled it to the kinetic law in order to find kinetic parameters independent of the reactor's geometry.

For solving the time-dependent mass balance, the whole system (reaction zone, recycle tank, and piping) was considered as a combination of a series of plug flow reactors (PFRs) and continuous stirred tanks under turbulent flow and resolved for the total organic carbon (TOC) concentration. Moreover, a Taguchi's robust design was used for estimating the signal-to-noise ratio for each operating parameter value (catalyst load and initial pH). These values were determined for obtaining the highest mineralization of acetaminophen regardless of the variability of solar radiation. After finding these photocatalytic kinetic constants, the performance of solar photoreactors at different scales in terms of the theoretical footprint and the total area to total treating volume ratio (A_T/V_T) was analyzed.

2. Results

2.1. Signal-to-Noise Ratios of Initial pH and Catalyst Load

Table 1 shows the TOC removal obtained after varying the initial pH, catalyst load, and the solar accumulated energy. The highest (53.49%) and lowest (5.39%) TOC removals were obtained with a pH of 5, and the performance was favored by a higher catalyst load (0.6 g L^{-1}). It was expected that acidic pHs enhanced the photocatalytic degradation for the case of the acetaminophen. Regarding the solar accumulated UV energy, the degradation increased with higher values because of a higher quantity of available photons. In this case, the increase of TOC removal concerning the cloudy days was higher than 50%. Whereas with a pH of 9, the variability of the TOC removals was smaller than the observed one with a pH of 5.

Table 1. Total organic carbon (TOC) removal for the solar photocatalytic process.

Initial pH	[TiO$_2$] (g L^{-1})	Accumulated UV Energy	
		19.14 W h m^{-2}	38.28 W h m^{-2}
5.0	0.3	5.39	10.09
	0.6	32.05	53.49
9.0	0.3	38.92	47.80
	0.6	29.92	37.07

Regarding the catalyst load, the best results were obtained with 0.6 g L^{-1} for initial pH of 5 and with 0.3 g L^{-1} for initial pH of 8. This result can suggest an interaction between the catalyst load and the initial pH of the slurries.

The initial pH affects the physicochemical properties of the catalyst, including the surface charge, the size of aggregates, and the position of the conductance and valence bands [27–29]. The reported point of zero charge (pH$_{zpc}$) for the TiO$_2$ Degussa is between 6 and 6.5 [29,30]. When the pH of the slurry is below the pH$_{zpc}$, the surface of TiO$_2$ acquires a positive charge and vice versa. Therefore, when the initial pH is 5, the stronger electrostatic forces can enhance the attachment of anionic species derived from the primary target molecule or its intermediates. This phenomenon favors the adsorption of these species and their latter oxidation [11,31,32].

However, Horst et al. [30] have shown that when the pH is very close to 6, the TiO$_2$ particles aggregate with hydrodynamic diameters larger than those found in much more alkaline suspensions (i.e., pH = 9). Similarly, Vanegas et al. [33] have proven that when the pH is near to 5, the agglomeration of titania is stronger than in the case of suspensions having pH values around 8. The agglomeration of TiO$_2$ particles reduces its useful area, which is believed to diminish the photocatalytic mineralization rates. As a result, due to the electrostatic forces and the agglomeration effects, the highest and lowest mineralization percentages of acetaminophen could have been obtained with the same value of initial pH (Table 1).

Regarding the catalyst load, when the initial pH was 5 and the TiO_2 concentration was 0.3 g L^{-1}, the availability of active sites might not have been enough for obtaining elevated TOC removal rates due to the agglomeration effect mentioned before. On the contrary, when the catalyst load was 0.6 g L^{-1}, the higher amount of TiO_2 particles could have overcome the limitation imposed by their agglomeration. In any case, when the initial pH is 5, the instability of the TiO_2 suspension may diminish the effectiveness of the photocatalytic process because that value is very close to its pH_{zpc}.

On the other hand, according to Table 1, high TOC removals could also be attained when the initial pH was set to 9. With this pH, the availability of OH^- ions in solution enhances. As a result, we can expect that TOC removal rates rise because, according to the general mechanism of the TiO_2-based photocatalysis [22], the holes of the valence band react with OH^- ions to generate $OH^\bullet$.

Nevertheless, as mentioned before, when the solution pH is above the pH_{zpc}, the catalyst surface charges negatively, and consequently the adsorption of OH^- ions becomes more difficult. Probably, that is why the highest TOC removal was not obtained when working with this initial pH. In this case, the effect of catalyst load was opposite to that obtained with an initial pH of 5. The highest and lowest TOC removals were attained with 0.3 and 0.6 g L^{-1} of TiO_2, respectively. One possible reason may be the clouding effect taking place in the slurry when the TiO_2 concentration was 0.6 g L^{-1}. This effect could have occurred when the pH was 9 due to the higher dispersion of catalytic particles [29,34] in the reactive media, which could block the photons' path inside the reactor. Therefore, with an initial pH of 9 and catalyst load of 0.3 g L^{-1}, the clouding effect could have been minimal, and higher TOC removals were obtained.

Table 2 shows the S/N ratios of the initial pH and catalyst load employed in these experiments. According to it, the performance of the photocatalytic system is more robust (a steadier response with high variability of the noise factor) when the initial pH is 9 (S/N = 31.33), and the TiO_2 load is 0.6 g L^{-1} (S/N = 31.01). Concerning to the catalyst load, this result is different to the estimated by the SFM approach in a previous work that used a solar CPC reactor and P25 as the catalyst [34]. However, the SFM calculation of the cited report did not include the effect of the initial pH nor the adsorption phenomenon. As mentioned above, there could be an interaction between the initial pH and the catalyst load, and the optimal values can differ depending on the substrate and other operating conditions. The same discussion can be applied to the effect of the initial pH. It is important to note that the pH affects the surface charge of the solid (as mentioned previously) and the attack orientation of the $OH^\bullet$ as well, which can influence the oxidation rate significantly [26].

Table 2. Signal-to-noise ratios of the initial pH and catalyst concentration.

Variable	Level	S/R
Initial pH	5	19.43
	9	31.33
[TiO_2]	0.3 g L^{-1}	19.46
	0.6 g L^{-1}	31.01

The most relevant result is the discrepancy with other studies with similar operating conditions (initial pH and catalyst load) but using a controlled or fixed amount of UV accumulated energy. Whereas in previous works [34–36], the reported values for catalyst loads were around 0.35 g L^{-1} (closer to the lower catalyst load used in this study), the recommended one in this study is 0.6 g L^{-1}. In fact, from Table 1, the TOC removal at pH of 9 was higher with 0.3 g L^{-1}, which is more consistent with the reported in the literature for this kind of reactor [35,37]. However, since the target of the Taguchi experimental design is finding suitable operating conditions for a robust operation (regardless to the variation of the solar radiation), the selection of 0.6 g L^{-1} is justified by a large number of active sites when the available UV photons are scarce (cloudy days) or particle agglomeration reduces the surface area due to the pH effect. Furthermore, the initial pH of 9 is far from the pH_{zpc} of the P25, and therefore, the apparent particle size becomes more stable, which improves the photocatalytic process.

Additionally, as the effects of the photolysis and the physical adsorption in all experiments were negligible (0.93%–1.17% of TOC removal). Therefore, it can be stated that the TOC removal can be attributed mainly to the photocatalytic oxidation process.

2.2. TOC Removal Modeling

The TOC removal was modeled by coupling the hydrodynamics, photonics, kinetics and mass balance in the photocatalytic reactor. The L-H parameters were obtained by fitting the experimental data to the mathematical model. The reaction time was standardized according to the commonly used t_{30W} expression, which is a normalization of the time that considers continuous irradiation of 30 W m^{-2} over the reactive zone [24,38,39].

The L-H parameters were estimated through linear regression of the reciprocal values of the initial rates and initial concentrations (initial rates law) according to the Equation (1). This equation is the reciprocal of the material balance expression associated with the batch-recirculating photocatalytic system. Figure 1 shows the fitted linear analysis for the initial reaction rates obtained with the three different TOC initial concentrations. This strategy allowed to find kinetic parameters independent of the radiation field, as described in Equations (2) and (3). In Equation (2), VRPA represents the integration of the LVRPA along the reactor volume, whereas the 1.2×10^4 factor was used as a conversion factor from ppm to mol L^{-1}.

$$\frac{1}{V_T \left(-\frac{dTOC}{dt_{30w}}\right)_{t=0}} = \frac{1}{k_T K_1 \int_{V_R} (LVRPA)^m dV_R} \left(\frac{1}{TOC_0}\right) + \frac{1}{k_T \int_{V_R} (LVRPA)^m dV_R} \quad (1)$$

$$k_T = \frac{1}{(1.2 \times 10^4)(1071.5)(VRPA)} = 7.59 \times 10^{-8} \text{ mol L}^{-1}\text{s}^{-1}\text{W}^{-0.5}\text{m}^{1.5} \quad (2)$$

$$K_1 = \frac{(1\,071.5)(1.2 \times 10^4)}{117\,401} = 109.52 \text{ L mol}^{-1} \quad (3)$$

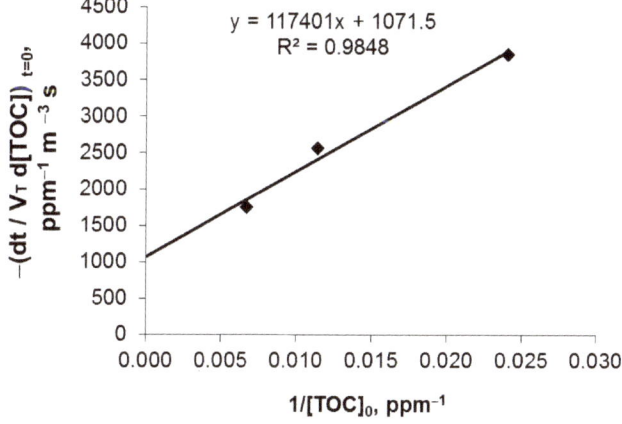

Figure 1. Linear regression for estimation of the Langmuir-Hinshelwood kinetic parameters.

Although the kinetics parameters were obtained with data of the bulk of the liquid-phase, they are still valid to represent this model because the external mass-transfer limitations between the bulk and the catalyst surface are negligible due to the significant mass-transport in the turbulent regime. As a result, the TOC found in the surroundings of the catalyst surface can be assumed as the same found in the bulk of the solution. Besides, there are no internal mass-transfer limitations because the catalyst (TiO$_2$ Degussa P25) is considered non-porous.

Furthermore, as it will be explained in Section 3.5, Equation (21) was integrated in order to obtain the TOC removal profile, as follows:

$$TOC_{r,\theta}^{out} = \exp\left[\ln\left(TOC^{in}\right) - \frac{K_1 k_T}{v_z(1 + K_1 TOC_0)} \int_0^L (LVRPA)_{r,\theta}^m dz\right] \quad (4)$$

where L is the total length of the reactor and m was taken as 0.5, which is a value suitable for geographical zones closer to the Earth's equator, where radiation intensities are high, and there is a good photon availability [22]. The boundary conditions to solve Equation (4) were

$$Z = 0, \; TOC = TOC^{in} \quad (5)$$

where TOC^{in} represents the inlet TOC concentration of the reactor in a given time. This model was employed to predict the TOC abatement of the contaminant according to the three different initial concentrations. The results are presented in the Figure 2, which reveals a good fit of the model to the experimental data.

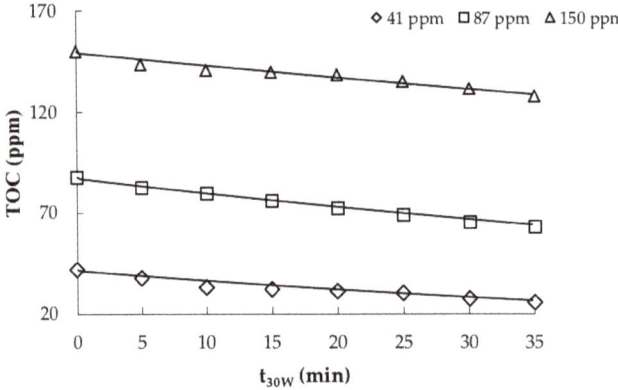

Figure 2. TOC photocatalytic removal. Model (solid line) vs. experimental data (markers).

From the obtained results, the kinetic parameters can be considered valid for the range of TOC initial concentrations (40–150 ppm) and an initial pH of 9. Although they were estimated with a single value of catalyst load, the model can be evaluated with different catalyst doses because the SFM consider the optical thickness and the scattering albedo, which are functions of the catalyst load. Nonetheless, the simulations were carried out under constant radiation flux of 30 W m^{-2}, which is considered as the average UV radiation flux received in a sunny day (10 a.m. to 2 p.m.) in northern latitudes. This value was selected considering previous works with solar radiation [24,34,39–41] and the difficulty of describing the solar radiation variability with the model used in this study.

2.3. Effect of Catalyst Load and Total Treated Volume on Plant Scaling-Up

The estimated kinetic parameters were used for simulating large-scale photocatalytic reactors in the TOC removal of acetaminophen. In order to compare the potential size of full-scale plants, the effect of the catalyst load and the total-pollutant volume on the mineralization concerning the A_T/V_T

ratio was studied. Figure 3 shows the TOC removal profiles in a system with a total volume of 5000 L, with two different catalyst loads (0.3 and 0.6 g L^{-1}).

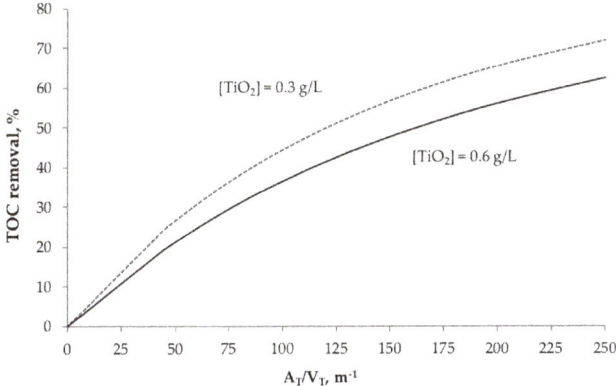

Figure 3. Effect of catalyst load on TOC removal, regarding the A_T/V_T ratio. (V_T = 5000 L, t_{30W} = 110 min, TOC_0 = 87.6 ppm).

The simulations were done under the same conditions described above (Equations (4), (5), and (8)–(11)) and the strategy is shown in Figure 4). The volume used of 5000 L represents the average-daily-wastewater volume generated in medium-sized Colombian hospitals or some industrial facilities.

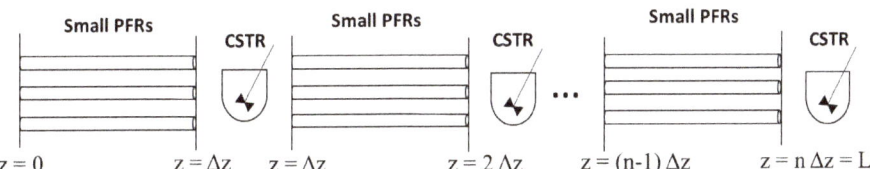

Figure 4. Strategy for modeling the photocatalytic reactor as a series of plug flow reactors (PFRs) and continuous stirring tanks (CSTRs).

The time t_{30W} was set at 110 min and the total area was estimated based on the footprint of a single CPC module (4.1 m^2, as seen in Figure 5). This footprint includes the area that would be covered by the whole CPC structure and the space between each module in a large-scale plant (30 cm of spacing lengthwise and crosswise).

The plot shows that the photocatalytic performance is better when using 0.3 g L^{-1} of the catalyst. This result was expected because the same photon-absorption model reported in ref. [34] was applied. As stated before, the optimal catalyst load in CPC reactors (regarding the LVRPA) was 0.3 g L^{-1}. As mentioned before, with an initial pH of 9 and 0.3 g L^{-1} of catalyst load, the best performance for TOC removal was obtained experimentally. Consequently, the simulations shown in Figure 3 are congruent with the experimental data. This optimal value for catalyst load is consistent with the results obtained in previous works, where solar CPCs of similar diameters were used under sunny weather conditions [35,37,39].

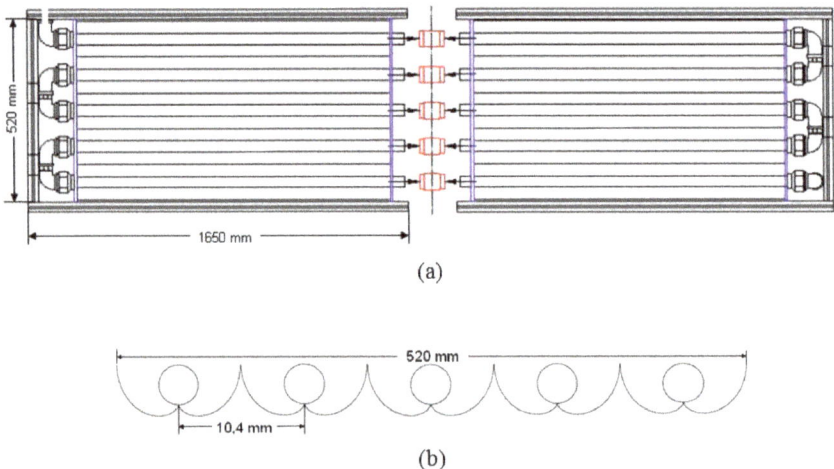

Figure 5. CPC reactor. (**a**) General scheme; (**b**) Tubes and reflectors (cross-sectional view).

The observed behavior in both plots (0.3 and 0.6 g L^{-1} in Figure 3) is similar for small ratios of A_T/V_T. However, better performances are shown for 0.3 g L^{-1} when the A_T/V_T ratio increases. For example, if a TOC removal of 50% is needed, a full-scale operation with 0.6 g L^{-1} of TiO$_2$ would require an A_T/V_T ratio of 175 m^{-1}; but if it operates with 0.3 g L^{-1} of TiO$_2$, the ratio would be 120 m^{-1}. The difference becomes more significant at higher TOC removals. This tendency can be explained due to the relative low mineralization rates that are usually obtained in photocatalytic processes. At small A_T/V_T ratios, there is no significant difference of TOC removal performance when they are low. When the A_T/V_T ratio increases, the residence time increases as well. Therefore, the conversion of the organic matter (via photocatalytic oxidation) is higher. Nevertheless, the mixing effect with the recycling-feeding tank (V_T) acts as a damping stage of the TOC removal process. Therefore, the overall degradation rate can become slower depending on the A_T/V_T.

From the above observation, the effect of the total treated volume (V_T) on the mineralization was evaluated as well. The photocatalytic abatement of acetaminophen with three different-contaminant volumes: 50, 500 and 5000 L (Figure 6) was simulated, which represent respectively the number of effluents that can be treated in the pilot, intermediate, and full-scale plants.

Figure 6 shows that the TOC removals for the 5000 L curve are much higher than the ones corresponding to the 50 and 500 L curves, whose behavior is very similar. These results show that the area (or CPC modules) needed for obtaining a specific TOC removal is not directly proportional to the total volume. For example, if a TOC removal of 30% is required when treating a 500 L effluent (containing acetaminophen), then the most appropriate A_T/V_T ratio would be 200 m^{-1}. In contrast, if the volume of the effluent is 5000 L, then this ratio is reduced to a third part approximately (60 m^{-1}). The difference becomes substantially higher as the required TOC removal increases. As a result, this information may be tremendously useful when scaling photocatalytic processes, as it could avoid unnecessarily monetary investment for the construction and operation of the reactors.

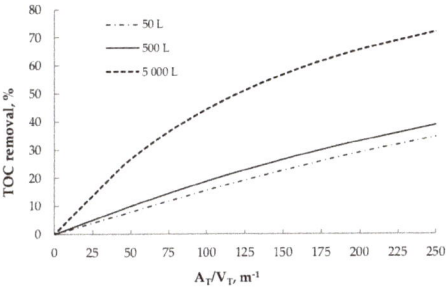

Figure 6. Effect of total volume on TOC removal, regarding the A_T/V_T ratio ([TiO_2] = 0.3 g L^{-1}, t_{30W} = 110 min, TOC_0 = 87.6 ppm).

In all the simulations, the flow rate was held constant at 30 L min^{-1}, which was the same used in the experimental runs carried out in the pilot-scale photoreactor. This supposition can be made because the CPC photoreactors are modular units that can be arranged in series. Therefore, a more extensive scale plant would only require a higher number of CPC reactors with the same size and operating conditions than the CPC used in the experimental pilot-scale plant. Nonetheless, larger volumes with the same flow rate yield higher residence times, which can improve the TOC removal as seen in Figure 6. Then, in order to scale-up and design full-scale plants with solar CPC photoreactors, it is not enough to estimate an ideal A_T/V_T ratio for attaining a given TOC removal. It requires a more in-depth analysis that can be done with the model presented here.

At large-scale, the effect of the initial concentration of acetaminophen was insignificant. Several simulations conducted with [TiO_2] = 0.6 g L^{-1}, V_T = 5000 L at 41.6 ppm, 87.6 ppm, and 149.8 ppm showed almost null differences in TOC removal (results not presented here).

3. Materials and Methods

3.1. Reagents and Chemicals

A stock solution of the contaminant was prepared with commercial liquid acetaminophen (Genfar®-Sanofi, Bogota, Colombia). TiO_2 Aeroxide P-25 (Evonik, Essen, Germany) was employed as the photocatalyst in all the experiments (primary particle size, ~21 nm by TEM; specific surface area 50 m^2 g^{-1} by BET; composition 80% anatase and 20% rutile by X-ray diffraction). The initial pH was adjusted with solutions of NaOH 0.1 N and HCl 0.1 N (Merck, Darmstadt, Germany).

3.2. Equipment

The experimental runs were carried out in the Solar Photocatalysis Laboratory at Universidad del Valle (Cali, Colombia—3°29′N latitude). Figure 5 exhibits a schematic representation of the CPC photoreactor used in this study. It consisted of 10 Duran glass tubes (1200 mm in length, 32 mm o.d., 1.4 mm wall thickness) that were placed upon a series of involutes made of aluminum (reflectance: ψ = 0.85), as seen in Figure 5b. The reactor was operated under a batch regime with recirculation, using a 40 L recycle feed tank and a centrifugal pump (0.5 HP of nominal power) that delivered 30.2 L min^{-1}. This experimental setup made it possible to keep the slurry (fluid and catalyst) saturated with oxygen, because whenever the slurry left the pipe, it was exposed to the surrounding air before entering the tank. The whole piping and accessories were made of PVC, 1 in. diameter.

The TOC concentration was followed with a TOC analyzer (Shimadzu 5050A, Sao Paulo, Brazil); whereas, the pH was measured with an Orion 4-Star pH-meter (Thermo Scientific-ARC Analisis, Bogota, Colombia). Additionally, the solar UV intensity and the corresponding accumulated energy in the 295-380 nm range were measured with a UV A+B radiometer (Solardetox-Acadus S50, Barcelona, Spain).

3.3. Experimental Design

Due to the weather variability (sunny or cloudy days in tropical regions), the photocatalytic mineralization of acetaminophen was evaluated with the Taguchi's experimental design [42,43]. This is a robust design that allows finding the most appropriate operational conditions that are insensitive to the noise (non-controllable factors), through the estimation of the signal-to-noise ratios (S/N) of the controllable variables. In this study, the initial pH and catalyst load were chosen as the controllable factors because they have been reported as two of the most influencing variables in the performance heterogeneous photocatalytic reactions [27,37,44]. Similarly, the accumulated solar UV energy is another important parameter, but it cannot be controlled since it depends on the geographical location, weather conditions and time of the day. As a result, the accumulated solar UV energy was considered as the noise factor. The corresponding signal-to-noise ratios (S/N) of each controllable variable were estimated with the "more is better" equation from the Taguchi's robust experimental design (Equation (6)) because our purpose was to maximize the mineralization of acetaminophen.

$$\frac{S}{N} = -10 \log \left(\frac{1}{n} \sum \frac{1}{Y_i^2} \right) \quad (6)$$

Regarding Equation (6), S/N stands for the signal-to-noise ratio of each level of the experimental factors while Y_i and n represent the percentage of mineralization and the number of experiments associated with each level. The TOC removal was calculated with the Equation (7).

$$\% \text{ Mineralization} = \frac{TOC_i - TOC_o}{TOC_i} \quad (7)$$

where TOC_i and TOC_o represent the TOC at the beginning and the end of each experimental run, respectively.

3.4. Procedure

The initial TOC concentration of acetaminophen was set to 40 ppm to simulate the strength of the wastewater generated in the washing containers and glass equipment from the Drugs Laboratory at the Universidad de Cartagena, Colombia. The initial pH was set to 5 and 9 for avoiding extreme conditions of acidity or alkalinity, which would require further amounts of reagents for neutralization.

The catalyst loads were 0.3 and 0.6 g L^{-1}, which are within the range reported in the literature [45,46]. In the first stage of the experimental runs, the samples from the reactor were taken at the beginning of the process and after the amount of UV energy reached 19.14 and 38.28 W h m^{-2}. These values represent the average quantity of accumulated solar-UV energy received in Cali during a 3-h period in cloudy and sunny days, respectively. Subsequently, the samples were taken and filtered using 0.45 µm membranes (Merck Millipore®, Cartagena, Colombia) for measuring the removal of TOC. Afterward, three different initial concentrations (40, 90 and 150 ppm of TOC) were considered for finding the kinetic parameters of the photocatalytic process. In this case, the initial pH and catalyst dosage were set to the values that exhibited the highest S/N ratio described in Section 3.3, and the reactor was operated until it reached 35 W h m^{-2} of solar UV accumulated energy. Here, the samples were also taken at the beginning and the end of the experiments, and each 5 W h m^{-2} of accumulated UV energy.

These runs were conducted under sunny weather conditions, and the flow rate was held at 30.2 L min^{-1} to ensure turbulent flow (Reynolds number = 19,420). In all cases, in order to achieve adsorption equilibrium, the slurry was recirculated for 20 min under dark conditions.

3.5. Modeling of the Solar CPC Photoreactor

The modeling approach consisted of coupling hydrodynamics with a photocatalytic kinetic model (including the LVRPA) in a time-dependent mass balance, as previously described in ref. [23].

The hydrodynamics was described by the following equations [24,47]:

$$\frac{v_z}{v_{z,\,max}} = \left(1 - \frac{r}{R}\right)^{1/n} \tag{8}$$

$$n = 0.41\sqrt{\frac{8}{f}} \tag{9}$$

$$v_{z,\,average} = \frac{Q}{\pi R^2} \tag{10}$$

$$\frac{v_{z,max}}{v_{z,\,average}} = \frac{(n+1)(2n+1)}{2n^2} \tag{11}$$

in which r is the radial coordinate, n is a hydrodynamic parameter, and f is the friction factor. Further, the local volumetric rate of photon absorption (LVRPA) was estimated with the Equation (12) [21,24,34], which is derived from the SFM and adapted for a cylindrical configuration. The central assumption of this model is that the scattering phenomena takes place along the six Cartesians coordinates, which reduces the complexity of solving the photonic balance within the photocatalytic reactor. According to the SFM, the LVRPA is

$$\text{LVRPA} = \frac{I_0}{\lambda_{\omega corr}\omega_{corr}(1-\gamma)}\left[\left(\omega_{corr} - 1 + \sqrt{1-\omega_{corr}^2}\right)e^{-r_p/\lambda_{\omega corr}} + \gamma\left(\omega_{corr} - 1 - \sqrt{1-\omega_{corr}^2}\right)e^{r_p/\lambda_{\omega corr}}\right] \tag{12}$$

where I_0 corresponds to the solar UV radiation flux that hits the reactor wall (either direct or diffuse radiation); whereas r_p is a parameter considered in the SFM which is associated to the photon's traveling path [21,24], and γ, ω_{corr}, $\lambda_{\omega corr}$ are defined as follows:

$$\omega_{corr} = \frac{b}{a} \tag{13}$$

$$a = 1 - \omega p_f - \frac{4\omega^2 p_s^2}{1 - \omega p_f - \omega p_b - 2\omega p_s} \tag{14}$$

$$b = \omega p_b + \frac{4\omega^2 p_s^2}{(1 - \omega p_f - \omega p_b - 2\omega p_s)} \tag{15}$$

$$\omega = \frac{\sigma}{\sigma + \kappa} \tag{16}$$

$$\gamma = \frac{1 - \sqrt{1-\omega_{corr}^2}}{1 + \sqrt{1-\omega_{corr}^2}} \exp(-2\tau_{app}) \tag{17}$$

$$\lambda_{\omega corr} = \frac{1}{a(\sigma + \kappa)c_{cat}\sqrt{1-\omega_{corr}^2}} \tag{18}$$

$$\tau_{app} = a\tau\sqrt{1-\omega_{corr}^2} \tag{19}$$

$$\tau = (\sigma + \kappa)c_{cat}\delta_{SFM} \tag{20}$$

The simulation of the radiant field and the calculation of the LVRPA were done in a Visual Basic routine that coupled the Ray Tracing technique with the SFM and a radiant emission model. This was previously reported in refs. [24,34,40]. As the LVRPA appears as the photonic contribution in the kinetic law, it is feasible to find kinetic parameters independent of the radiation field.

The mass balance was solved in terms of the TOC and was coupled to a hydrodynamic model for turbulent regime [24,47]. Moreover, the kinetics contribution was described with a Langmuir-Hinshelwood (L-H) equation that had an explicit dependence on the LVRPA, as shown on the right-hand side of Equation (21).

The entire reactor was divided into a large number (2500) of plug flow reactors (PFR) of length L, each one of them associated to the (r,θ) coordinates of the cross-sectional area. A higher number of PFRs would represent a significant increase in the computing time (as observed in previous simulations not shown in this study), without a visible improvement in the accuracy of the model. In that case, the mass balance can be described by the following equation:

$$Q\frac{dTOC_{r,\theta}}{dV_R} = -\frac{k_t K_1 TOC_{r,\theta}}{(1+K_1 TOC_0)}(LVRPA)_{r,\theta}^m \quad (21)$$

where k_t and K_1 represent the kinetic and binding constants, respectively. In each PFR, the flow rate Q was equivalent to the product of the cross-section area and the average axial velocity ($A_R\, v_z$). Besides, dV_R in Equation (16) could be replaced by $A_R\, dz$, so that we obtained a differential equation in function just of the axial direction z. Thereby, it was only necessary to estimate the average axial velocities profile in terms of the radial coordinate (r). Although the first option for modeling this kind of photoreactor is to consider it as a PFR, this is not entirely accurate due to the turbulent regime of the system. The mixing of the streamlines does not allow to find a well-defined velocity profile; therefore, since the concentration depends on the velocity due to the convective effects, the turbulent regime must be considered for the mass balance of the reactor. The CSTR provides a simple way for modeling this part of the mass transfer phenomenon with more accuracy than the PFR alone.

Considering the above assumption, each PFR was divided into a series of small reactors with a length of Δz. In every simulation step, Equation (21) was solved for each small reactor, starting from the plane z = 0 down to z = Δz. Then, in order to consider the mixing effect of the turbulent regime, the TOC profile was averaged in the z = Δz plane. This averaging step, as shown in Equation (17), was intended for assuming that a continuous stirring tank (CSTR) was in that position.

$$TOC_{average}^{out} = \frac{\int_0^{2\pi}\int_0^{2R} r v_z TOC_{r,\theta}^{out} dr d\theta}{Q} \quad (22)$$

Afterward, this average was taken as the inlet concentration for the next PFR reactor located from z = Δz to z = 2Δz, being Δz = L/100. Subsequently, Equation (21) was solved for each reactor found in the (z = Δz, z = 2Δz) interval, and a new CSTR was virtually placed in z = 2Δz. Finally, these steps were repeated successively until the total length of the reactor was covered (z = L). This modeling approach is described graphically in the Figure 4, in which n corresponds to the number of divisions of the total length (L).

The time dependence of the photocatalytic process was treated as a step dependence, which is associated with the number of passes (n_{pass}) that the slurry has in the reactor. This strategy has been used several times for modeling photocatalytic recirculation systems [24,37,48].

$$n_{pass} = \frac{Qt_{30W}}{V_R} \quad (23)$$

The change in TOC concentration per each pass was estimated as follows:

$$TOC_{i+1}^{in} = \frac{TOC_i^{in}(V_T - V_R) + TOC_i^{out}V_R}{V_T} \quad (24)$$

4. Conclusions

The Taguchi experimental design was applied for analyzing the TOC removal of commercial acetaminophen in a solar CPC photocatalytic reactor. It showed that the most favorable conditions for a robust operation were an initial pH of 9 and a catalyst load of 0.6 g L^{-1}. Although the results differ from the reported studies with similar conditions, the variation of the solar radiation and the interaction of the pH with the catalyst load are the reasons for this discrepancy. On the other hand, the kinetic parameters obtained through the mathematical model proposed in this work (k_T = 7.5874 × 10^{-8} mol L^{-1} s^{-1} $W^{-0.5}$ $m^{1.5}$ and K_1 = 109.52 L mol^{-1}) can be used for scaling purposes since the model had a specific contribution of the photonic absorption. Furthermore, that large-scale plants require smaller ratios of A_T/V_T when compared with intermediate and pilot-scale schemes. This result is reasonable because the higher the scale, the higher residence times, and therefore, the conversion is enhanced. Therefore, in order to save monetary resources, a careful analysis based on these results should be made before deciding to scale photocatalytic reactors.

Author Contributions: Déyler Castilla-Caballero and José Colina-Márquez analyzed the data and run the simulations of the kinetic model; Fiderman Machuca-Martínez supplied the reactants and the equipment for the experimental runs; and Ciro Bustillo-Lecompte collaborated with the paper writing and editing.

Acknowledgments: The authors acknowledge Chemical engineers Juan Cohen, Leonardo Narváez, Rodinson Arrieta, and María Guerra for carrying out the experimental tests and simulations. Writing assistance provided by Helen Burnham from the British Council is much appreciated. The authors also thank Colciencias for funding this research (Grant No. 110752128546) and their Ph.D. studies.

Conflicts of Interest: The authors declare no conflict of interest.

References

1. Margot, J.; Rossi, L.; Barry, D.A.; Holliger, C. A review of the fate of micropollutants in wastewater treatment plants. *Wiley Interdiscip. Rev. Water* **2015**, *2*, 457–487. [CrossRef]
2. Boethling, R.; Fenner, K.; Howard, P.; Klečka, G.; Madsen, T.; Snape, J.R.; Whelan, M.J. Environmental persistence of organic pollutants: Guidance for development and review of POP risk profiles. *Integr. Environ. Assess. Manag.* **2009**, *5*, 539–556. [CrossRef] [PubMed]
3. Liu, G.; Yang, Z.; Tang, Y.; Ulgiati, S. Spatial correlation model of economy-energy-pollution interactions: The role of river water as a link between production sites and urban areas. *Renew. Sustain. Energy Rev.* **2017**, *69*, 1018–1028. [CrossRef]
4. Sayal, A.; Amjad, S.; Bilal, M.; Pervez, A.; Mahmood, Q.; Afridi, M. Industrial Water Contamination and Health Impacts: An Economic Perspective. *Pol. J. Environ. Stud.* **2016**, *25*, 765–775. [CrossRef]
5. Ros, O.; Izaguirre, J.K.; Olivares, M.; Bizarro, C.; Ortiz-Zarragoitia, M.; Cajaraville, M.P.; Etxebarria, N.; Prieto, A.; Vallejo, A. Determination of endocrine disrupting compounds and their metabolites in fish bile. *Sci. Total Environ.* **2015**, *536*, 261–267. [CrossRef] [PubMed]
6. Heberer, T. Occurrence, fate, and removal of pharmaceutical residues in the aquatic environment: A review of recent research data. *Toxicol. Lett.* **2002**, *131*, 5–17. [CrossRef]
7. Rodrigues, R.M.; Sachinidis, A.; de Boe, V.; Rogiers, V.; Vanhaecke, T.; de Kock, J. Identification of potential biomarkers of hepatitis B-induced acute liver failure using hepatic cells derived from human skin precursors. *Toxicol. In Vitro* **2015**, *29*, 1231–1239. [CrossRef] [PubMed]
8. Wang, Y.; Jiang, Y.; Fan, X.; Tan, H.; Zeng, H.; Wang, Y.; Chen, P.; Huang, M.; Bi, H. Hepato-protective effect of resveratrol against acetaminophen-induced liver injury is associated with inhibition of CYP-mediated bioactivation and regulation of SIRT1-p53 signaling pathways. *Toxicol. Lett.* **2015**, *236*, 82–89. [CrossRef] [PubMed]
9. SanJuan-Reyes, N.; Gómez-Oliván, L.M.; Galar-Martínez, M.; García-Medina, S.; Islas-Flores, H.; González-González, E.D.; Cardoso-Vera, J.D.; Jiménez-Vargas, J.M. NSAID-manufacturing plant effluent induces geno- and cytotoxicity in common carp (*Cyprinus carpio*). *Sci. Total Environ.* **2015**, *530–531*, 1–10. [CrossRef] [PubMed]
10. Ribas, J.L.C.; da Silva, C.A.; de Andrade, L.; Galvan, G.L.; Cestari, M.M.; Trindade, E.S.; Zampronio, A.R.; de Assis, H.C.S. Effects of anti-inflammatory drugs in primary kidney cell culture of a freshwater fish. *Fish Shellfish Immunol.* **2014**, *40*, 296–303. [CrossRef] [PubMed]
11. Grčić, I.; Puma, G.L. Photocatalytic degradation of water contaminants in multiple photoreactors and evaluation of reaction kinetic constants independent of photon absorption, irradiance, reactor geometry, and hydrodynamics. *Environ. Sci. Technol.* **2013**, *47*, 13702–13711. [CrossRef] [PubMed]
12. Belver, C.; Bedia, J.; Rodriguez, J.J. Zr-doped TiO_2 supported on delaminated clay materials for solar photocatalytic treatment of emerging pollutants. *J. Hazard. Mater.* **2017**, *322*, 233–242. [CrossRef] [PubMed]

13. Fagan, R.; McCormack, D.E.; Dionysiou, D.D.; Pillai, S.C. A review of solar and visible light active TiO$_2$ photocatalysis for treating bacteria, cyanotoxins and contaminants of emerging concern. *Mater. Sci. Semicond. Process.* **2016**, *42*, 2–14. [CrossRef]
14. Miranda-García, N.; Suárez, S.; Maldonado, M.I.; Malato, S.; Sánchez, B. Regeneration approaches for TiO$_2$ immobilized photocatalyst used in the elimination of emerging contaminants in water. *Catal. Today* **2014**, *230*, 27–34. [CrossRef]
15. Turchi, C.; Ollis, D. Photocatalytic degradation of organic water contaminants: Mechanisms involving hydroxyl radical attack. *J. Catal.* **1990**, *122*, 178–192. [CrossRef]
16. Alfano, O.M.; Cabrera, I.; Cassano, A.E. Photocatalytic Reactions Involving Hydroxyl Radical Attack I Reaction Kinetics Formulation with Explicit Photon Absorption Effects. *J. Catal.* **1997**, *172*, 370–379. [CrossRef]
17. Romero, R.L.; Alfano, A.O.M.; Cassano, A.E. Cylindrical Photocatalytic Reactors. Radiation Absorption and Scattering Effects Produced by Suspended Fine Particles in an Annular Space. *Ind. Chem. Eng. Res.* **1997**, *36*, 3094–3109. [CrossRef]
18. Satuf, M.L.; Brandi, R.J.; Cassano, A.E.; Alfano, O.M. Modeling of a flat plate, slurry reactor for the photocatalytic degradation of 4-chlorophenol. *Int. J. Chem. React. Eng.* **2007**, *5*. [CrossRef]
19. Pasquali, M.; Santarelli, F.; Porter, J.F.; Yue, P.-L. Radiative transfer in photocatalytic systems. *AIChE J.* **1996**, *42*, 532–537. [CrossRef]
20. Spadoni, G.; Bandini, E.; Santarelli, F. Scattering effects in photosensitized reactions. *Chem. Eng. Sci.* **1978**, *33*, 517–524. [CrossRef]
21. Brucato, A.; Grisafi, C.; Montante, G.; Rizzuti, G.; Vella, G. Estimating radiant fields in flat heterogeneous photoreactors by the six-flux model. *AIChE J.* **2006**, *52*, 3882–3890. [CrossRef]
22. Mills, A.; O'Rourke, C.; Moore, K. Powder semiconductor photocatalysis in aqueous solution: An overview of kinetics-based reaction mechanisms. *J. Photochem. Photobiol. A Chem.* **2015**, *310*, 66–105. [CrossRef]
23. Boyjoo, Y.; Ang, M.; Pareek, V. Some aspects of photocatalytic reactor modeling using computational fluid dynamics. *Chem. Eng. Sci.* **2013**, *101*, 764–784. [CrossRef]
24. Colina-Márquez, J.; Machuca-Martínez, F.; Puma, G.L. Photocatalytic mineralization of commercial herbicides in a pilot-scale solar CPC reactor: Photoreactor modeling and reaction kinetics constants independent of radiation field. *Environ. Sci. Technol.* **2009**, *43*, 8953–8960. [CrossRef] [PubMed]
25. Khorasanizadeh, H.; Mohammadi, K. Diffuse solar radiation on a horizontal surface: Reviewing and categorizing the empirical models. *Renew. Sustain. Energy Rev.* **2016**, *53*, 338–362. [CrossRef]
26. Colina-Márquez, J.; Machuca-Martínez, F.; Puma, G. Modeling the Photocatalytic Mineralization in Water of Commercial Formulation of Estrogens 17-β Estradiol (E2) and Nomegestrol Acetate in Contraceptive Pills in a Solar Powered Compound Parabolic Collector. *Molecules* **2015**, *20*, 13354–13373. [CrossRef] [PubMed]
27. Basha, S.; Keane, D.; Nolan, K.; Oelgemöller, M.; Lawler, J.; Tobin, J.M.; Morrissey, A. UV-induced photocatalytic degradation of aqueous acetaminophen: The role of adsorption and reaction kinetics. *Environ. Sci. Pollut. Res.* **2014**, *22*, 2219–2230. [CrossRef] [PubMed]
28. Ramirez-Garcia, S.; Chen, L.; Morris, M.A.; Dawson, K.A. A new methodology for studying nanoparticle interactions in biological systems: Dispersing Titania in biocompatible media using chemical stabilisers. *Nanoscale* **2011**, *3*, 4617–4624. [CrossRef] [PubMed]
29. Palma-Goyes, R.E.; Silva-Agredo, J.; González, I.; Torres-Palma, R.A. Comparative degradation of indigo carmine by electrochemical oxidation and advanced oxidation processes. *Electrochim. Acta* **2014**, *140*, 427–433. [CrossRef]
30. Horst, A.M.; Ji, Z.; Holden, P.A. Nanoparticle dispersion in environmentally relevant culture media: A TiO$_2$ case study and considerations for a general approach. *J. Nanopart. Res.* **2012**, *14*, 1014. [CrossRef]
31. Segalin, J.; Sirtori, C.; Jank, L.; Lima, M.F.S.; Livotto, P.R.; Machado, T.C.; Lansarin, M.A.; Pizzolato, T.M. Identification of transformation products of rosuvastatin in water during ZnO photocatalytic degradation through the use of associated LC-QTOF-MS to computational chemistry. *J. Hazard. Mater.* **2015**, *299*, 78–85. [CrossRef] [PubMed]
32. Długosz, M.; Żmudzki, P.; Kwiecień, A.; Szczubiałka, K.; Krzek, J.; Nowakowska, M. Photocatalytic degradation of sulfamethoxazole in aqueous solution using a floating TiO$_2$-expanded perlite photocatalyst. *J. Hazard. Mater.* **2015**, *298*, 146–153. [CrossRef] [PubMed]

33. Vanegas, M.E.; Vázquez, V.; Moscoso, D.; Cruzat, C. Síntesis y caracterización de nanopartículas magnéticas del tipo Fe_3O_4/TiO_2, efecto del pH en la dispersión y estabilización en soluciones acuosas. *Maskana* **2014**, *5*, 43–55.
34. Colina-Márquez, J.; Díaz, D.; Rendón, A.; López-Vásquez, A.; Machuca-Martínez, F. Photocatalytic treatment of a dye polluted industrial effluent with a solar pilot-scale CPC reactor. *J. Adv. Oxid. Technol.* **2009**, *12*, 93–99. [CrossRef]
35. Colina-Márquez, J.; Machuca-Martínez, F.; Li Puma, G. Radiation absorption and optimization of solar photocatalytic reactors for environmental applications. *Environ. Sci. Technol.* **2010**, *44*, 5112–5120. [CrossRef] [PubMed]
36. López-Vásquez, A.; Ortiz, E.; Arias, F.; Colina-Márquez, J.A.; Machuca, F. Photocatalytic decolorization of methylene blue with two photoreactors. *J. Adv. Oxid. Technol.* **2008**, *11*, 33–48. [CrossRef]
37. Mueses, M.A.; Machuca-Martinez, F.; Li Puma, G. Effective quantum yield and reaction rate model for evaluation of photocatalytic degradation of water contaminants in heterogeneous pilot-scale solar photoreactors. *Chem. Eng. J.* **2013**, *215–216*, 937–947. [CrossRef]
38. Fenoll, J.; Flores, P.; Martínez, C.; Navarro, S. Photodegradation of eight miscellaneous pesticides in drinking water after treatment with semiconductor materials under sunlight at pilot plant scale. *Chem. Eng. J.* **2012**, *1*, 2204–2206. [CrossRef]
39. Prieto-Rodriguez, L.; Miralles-Cuevas, S.; Oller, I.; Agüera, A.; Puma, G.L.; Malato, S. Treatment of emerging contaminants in wastewater treatment plants (WWTP) effluents by solar photocatalysis using low TiO_2 concentrations. *J. Hazard. Mater.* **2012**, *211–212*, 131–137. [CrossRef] [PubMed]
40. Colina-Márquez, J.; Lopez-Vasquez, A.; Machuca-Martínez, F. Modeling of direct solar radiation in a compound parabolic collector (CPC) with the ray tracing technique. *Dyna-Colombia* **2010**, *77*, 132–140.
41. Nasirian, M.; Lin, Y.P.; Bustillo-Lecompte, C.F.; Mehrvar, M. Enhancement of photocatalytic activity of titanium dioxide using non-metal doping methods under visible light: A review. *Int. J. Environ. Sci. Technol.* **2017**, 1–24. [CrossRef]
42. Twu, M.J.; Chiou, A.H.; Hu, C.C.; Hsu, C.Y.; Kuo, C.G. Properties of TiO_2 films deposited on flexible substrates using direct current magnetron sputtering and using high power impulse magnetron sputtering. *Polym. Degrad. Stab.* **2015**, *117*, 1–7. [CrossRef]
43. Nikazar, M.; Gorji, L.M.; Shojae, S.; Keynejad, K.; Haghighaty, A.H.; Jalili, F.; Mirzahosseini, A.R.H. Removal of Methyl Tertiary-Butyl Ether (MTBE) from Aqueous Solution Using Sunlight and Nano TiO_2. *Energy Sources Part A Recover. Util. Environ. Eff.* **2014**, *36*, 2305–2311. [CrossRef]
44. Ghodbane, H.; Hamdaoui, O.; Vandamme, J.; van Durme, J.; Vanraes, P.; Leys, C.; Nikiforov, A.Y. Degradation of AB25 dye in liquid medium by atmospheric pressure non-thermal plasma and plasma combination with photocatalyst TiO_2. *Open Chem.* **2015**, *13*, 325–331. [CrossRef]
45. Borges, M.; García, D.; Hernández, T.; Ruiz-Morales, J.; Esparza, P. Supported Photocatalyst for Removal of Emerging Contaminants from Wastewater in a Continuous Packed-Bed Photoreactor Configuration. *Catalysts* **2015**, *5*, 77–87. [CrossRef]
46. Escapa, C.; Coimbra, R.N.; Paniagua, S.; García, A.I.; Otero, M. Nutrients and pharmaceuticals removal from wastewater by culture and harvesting of Chlorella sorokiniana. *Bioresour. Technol.* **2015**, *185*, 276–284. [CrossRef] [PubMed]
47. Bird, R.B.; Stewart, W.E.; Lightfoot, E.N. *Transport Phenomena*; Wiley: Hoboken, NJ, USA, 2007.
48. Puma, G.L.; Khor, J.; Brucato, A. Modeling of an annular photocatalytic reactor for water purification: Oxidation of pesticides. *Environ. Sci. Technol.* **2004**, *38*, 3737–3745. [CrossRef] [PubMed]

© 2018 by the authors. Licensee MDPI, Basel, Switzerland. This article is an open access article distributed under the terms and conditions of the Creative Commons Attribution (CC BY) license (http://creativecommons.org/licenses/by/4.0/).

Article

The CoAlCeO Mixed Oxide: An Alternative to Palladium-Based Catalysts for Total Oxidation of Industrial VOCs

Julien Brunet [1,*], Eric Genty [1,2], Cédric Barroo [2], Fabrice Cazier [3], Christophe Poupin [1], Stéphane Siffert [1], Diane Thomas [4], Guy De Weireld [4], Thierry Visart de Bocarmé [2] and Renaud Cousin [1,*]

[1] Unité de Chimie Environnementale et Interactions sur le Vivant, Université du Littoral Côte d'Opale, MREI1—145 Avenue Maurice Schumann, 59140 Dunkerque, France; eric.genty@ulb.ac.be (E.G.); Christophe.Poupin@univ-littoral.fr (C.P.); stephane.siffert@univ-littoral.fr (S.S.)
[2] Chemical Physics of Materials and Catalysis, Université Libre de Bruxelles, Faculty of Sciences, Campus Plaine CP 243, 1050 Brussels, Belgium; cbarroo@ulb.ac.be (C.B.); Thierry.Visart.de.Bocarme@ulb.ac.be (T.V.d.B.)
[3] Centre Commun de Mesures, Université du Littoral Côte d'Opale, MREI1—145 Avenue Maurice Schumann, 59140 Dunkerque, France; cazier@univ-littoral.fr
[4] Faculté Polytechnique de Mons, Université de Mons, 20 Place du Parc, B-7000 Mons, Belgium; Diane.THOMAS@umons.ac.be (D.T.); Guy.DEWEIRELD@umons.ac.be (G.D.W.)
* Correspondence: julien.brunet@univ-littoral.fr (J.B.); renaud.cousin@univ-littoral.fr (R.C.)

Received: 15 December 2017; Accepted: 1 February 2018; Published: 6 February 2018

Abstract: Catalytic total oxidation is an effective technique for the treatment of industrial VOCs principally resulting from industrial processes using solvents, and usually containing mono-aromatics (BTEX) and oxygenated compounds (acetone, ethanol, butanone). The catalytic total oxidation of VOCs on noble metal materials is effective. However, the cost of catalysts is a main obstacle for the industrial application of these VOC removal processes. Therefore, the aim of this work is to propose an alternative material to palladium-based catalysts (which are suitable for VOCs' total oxidation): a mixed oxide synthesized in the hydrotalcite way, namely CoAlCeO. This material was compared to four catalytic materials containing palladium, selected according to the literature: Pd/α-Al_2O_3, Pd/HY, Pd/CeO_2 and $Pd/\gamma Al_2O_3$. These materials have been studied for the total oxidation of toluene, butanone, and VOCs mixtures. Catalysts' performances were compared, taking into account the oxidation byproducts emitted from the process. This work highlight that the CoAlCeO catalyst presents better efficiency than Pd-based materials for the total oxidation of a VOCs mixture.

Keywords: mixed oxide catalyst; VOCs; byproducts; BTEX; catalytic total oxidation; CoAlCeO

1. Introduction

Volatile organic compounds (VOCs) are known as one of the major contributors to atmospheric pollution. Their anthropic release is particularly significant in industrialized areas and has noxious consequences for health, environment, and construction materials. The majority of VOC emissions originates from solvents used in several industrial sectors (paints, varnishes, lacquers, inks, adhesives, glues, etc.). A significant part of these solvents is composed of mono-aromatic compounds, particularly BTEX (benzene, toluene, ethylbenzene, and xylenes). The other part of these solvents is represented by oxygenated compounds, such as butanone (or methyl ethyl ketone (MEK)) which is a common oxygenated solvent, since it is a less toxic substitute for alcohols (methanol, ethanol) and acetone. Butanone and BTEX are also widely used in mixtures as solvent for lacquers, inks, and coatings for application to metal surfaces.

An effective method for the treatment of industrial VOC emissions is their catalytic oxidation which is a cost-effective and an ecological alternative to thermal oxidation, with similar efficiency. Studies of the BTEX oxidation, especially toluene, have been widely reported in the literature [1–7]. Oxidation of butanone is also well known in the literature, although the number of studies is more restricted [7–12]. However, studies of the catalytic oxidation of BTEX and oxygenated compounds in mixture are very limited, even though this is an important step towards industrial applications. Concerning the oxidation of VOCs mixture, the authors generally observe inhibition phenomena [7,13–18] that are either caused by competition between molecules, or at the level of the adsorption step to the surface of the catalyst during the mechanism of oxidation (a reaction with chemisorbed oxygen or lattice oxygen). These interactions depend on the VOC conformation and polarity, but also on the nature of the catalytic material. Furthermore, some molecules can react directly in the gas phase in contact with lattice oxygen, while others must be adsorbed to the catalyst surface before being oxidized. Oxidation of the latter will be strongly inhibited in favor of the former due to more favorable kinetics. Similarly, the oxidation of several compounds that need to be adsorbed on the catalyst will lead to adsorption competition.

In this work, a new and original approach to the total oxidation of VOCs is presented. Indeed, the catalytic oxidation of toluene and butanone alone or in mixture is presented for several materials, with a focus on the formed byproducts. Toluene and butanone have been chosen in accordance with effluents from industrial processes using lacquers, inks, and varnishes for application to metal surfaces. For this, various materials had been screened in order to find the most suitable catalyst for the catalytic treatment of VOCs from these industries. Firstly, palladium-based catalysts have been selected: Pd/HY, Pd/CeO$_2$, Pd/α-Al$_2$O$_3$, and a commercial Pd/γ-Al$_2$O$_3$. Palladium generally shows an equivalent or superior activity to platinum for the total oxidation of aromatic compounds [15,19,20]. Moreover, Pd is more resistant than Pt to sintering, the formation of volatile metal species, and poisoning by chlorine, water, and carbon monoxide [16,21,22]. Thus, these materials are particularly suited for the oxidation of VOC, especially BTEX. They have already been presented in a previous study on the total oxidation of toluene [3]. Secondly, as an alternative to the use of precious metals-based materials, our research group has conducted several studies to develop efficient transition metal oxides for environmental catalytic applications. Indeed, the catalyst cost is an important limitation for the industrial application of VOC catalytic oxidation processes. These studies led to the development of a Co–Al–Ce mixed oxide, denoted CoAlCeO, which reveals promising results concerning the oxidation of toluene [23]. Therefore, in this framework, these materials were compared for the total oxidation of toluene, butanone, and finally VOC mixtures (MEK/toluene and industrial mixtures).

2. Results

2.1. Catalysts Characterization

The catalysts used in this study were first characterized to obtain information on the specific surface area, the Pd content, the dispersion and particle size, as well as the molar ratio in the case of the CoAlCeO catalyst. These details are reported in Table 1. Furthermore, the X-ray diffraction patterns of the different materials are presented in Figure 1.

Table 1. Characterization of the studied catalytic materials.

	Pd/α-Al$_2$O$_3$	Pd/HY	Pd/CeO$_2$	Pd/γ-Al$_2$O$_3$	CoAlCeO
Specific surface area (m$^2 \cdot$g^{-1})	1	900	93	252	108
Palladium content (wt %)	0.48	0.46	0.40	0.40	N.A.
Palladium dispersion (%)	13.8	62.7	34.5	26.4	N.A.
Palladium particle size (nm)	8.1	1.8	3.2	4.3	N.A.
Co/Al/Ce Experimental molar ratio	N.A.	N.A.	N.A.	N.A.	6/1.1/0.8

Regarding palladium-based materials, different diffractograms show only the presence of the support, a sign that no supported palladium phase was detected. Nevertheless, the elementary analysis confirms the presence of palladium in several materials with relatively good deposition rates: this indicates that the palladium is deposited in the form of small nanoparticles whose size is lower than the detection limit of the diffractometer. This hypothesis was confirmed by hydrogen chemisorption measurements that indicate particle sizes between 1.8 and 8.1 nm. Concerning the $Co_6Al_{1.2}Ce_{0.8}O_x$ material, denominated CoAlCeO, the diffractogram reveals X-ray patterns characteristic of spinel-type oxides $M^{II}M^{III}_2O_4$ [24–26] corresponding to Co_3O_4, $CoAl_2O_4$ or Co_2AlO_4. In addition to the spinel phase, the diffraction pattern also showed the presence of a ceria phase (CeO_2) (Figure 1).

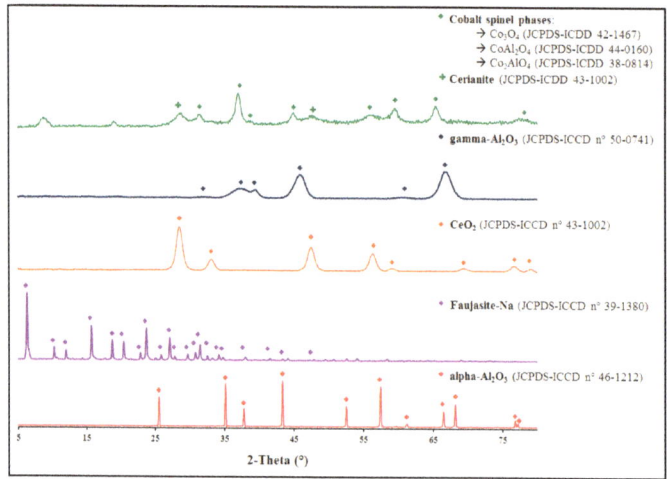

Figure 1. X-ray diffraction patterns of the catalysts.

2.2. Toluene Oxidation

As a first set of experiments, the materials of interest were used to investigate the total oxidation of toluene, and the light-off curves obtained are shown in Figure 2 for all materials.

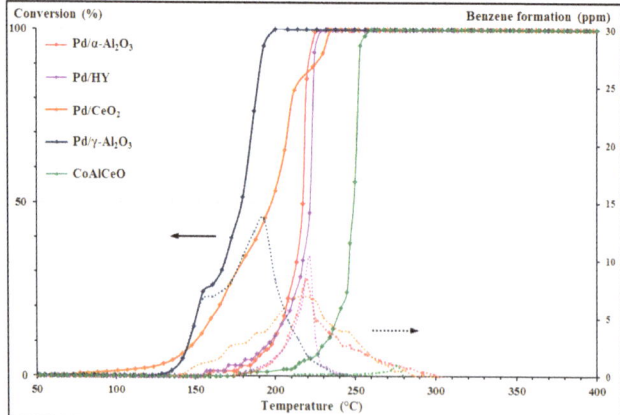

Figure 2. Toluene conversion (solid arrow and lines) and benzene formation (dashed arrow and lines) versus temperature for each catalyst.

The light-off curves of Pd-based catalysts have been presented in a previous study (Pd/α-Al$_2$O$_3$, Pd/HY, Pd/CeO$_2$ and Pd/γ-Al$_2$O$_3$) [3]. T_{50} and T_{100} are defined as the temperature when 50% and 100% conversion, respectively, was observed. These values of T_{50} and T_{100} for toluene oxidation are reported in Table 2. The test conducted with pure SiC shows only a slight toluene conversion from 360 °C, with a maximum of 5% at 400 °C. This can be explained by a thermal decomposition. Indeed, the test performed under the same conditions with the empty reactor leads to the same result. The results show that the mixed oxide has a lower performance than palladium-based catalysts. However, with a T_{50} value of 249 °C, the CoAlCeO material shows excellent activity for the toluene total oxidation. Moreover, in view of the activity values (A) in Table 2, it is possible to observe that the mixed oxide has a similar activity to that of Pd/CeO$_2$ and Pd/γ-Al$_2$O$_3$ catalysts, highlighting the possibility of such catalysts replacing Pd-based materials in the VOCs removal processes.

Table 2. T_{50} and T_{100} of catalysts for toluene catalytic oxidation and parameters characterizing the emission profiles of benzene.

Catalyst	T_{50} (°C)	T_{100} (°C)	A (mol/(m^2·h))	Q_{max} (ppm)	T_f (°C)	P (°C)
Pd/α-Al$_2$O$_3$	218	233	2.84 × 10^{-7}	8	300	67
Pd/HY	222	242	3.18 × 10^{-10}	10	249	7
Pd/CeO$_2$	197	235	3.34 × 10^{-9}	7	298	63
Pd/γ-Al$_2$O$_3$	179	200	1.27 × 10^{-9}	14	244	44
CoAlCeO	249	260	2.45 × 10^{-9}	0.9	280	20

As for palladium-based materials, the CoAlCeO mixed oxide was investigated regarding its properties to form byproducts, in particular benzene. In addition to the toluene conversion, Figure 2 shows the benzene emission profiles (as dashed lines) as a function of toluene conversion (solid lines) and temperature. The results show that the four palladium-based catalysts present similar emission profiles, with a maximum value at around 10 ppm. By contrast, the emission profile of the CoAlCeO mixed oxide revealed a maximum value of less than 1 ppm, namely an order of magnitude lower as compared to the other catalysts. In order to compare the benzene emissions of each material, the methodology presented in a previous study has been used [3], where the emission profiles are characterized by seven parameters. However, only the most relevant parameters are presented in this paper:

Q_{max}: maximum quantity emitted observed of the considered byproduct (ppm);
T_f: temperature at which the byproduct is totally oxidized (°C);
P: persistence of byproduct, corresponding to the difference between T_f and T_{100} (°C).

Table 2 reports the values of Q_{max}, T_f and P. Regarding Pd/α-Al$_2$O$_3$, Pd/HY, Pd/CeO$_2$, and Pd/γ-Al$_2$O$_3$ catalysts, the Q_{max} values present a maximum amount of emitted benzene between 7 and 14 ppm. These values are relatively low as compared to the concentration of oxidized toluene. However, by taking into account the toxicity of this compound and its strict regulations, these values remain important. On the contrary, the CoAlCeO catalyst presents the lowest emissions with a Q_{max} value of 0.9 ppm. Concerning the P values, they are positive (on average 40 °C), which clearly indicates that benzene is not fully oxidized when the toluene total oxidation (T_{100}) is reached. Therefore, this shows that benzene is a crucial limitation to the catalytic process of toluene oxidation: it is then necessary to consider this byproduct in order to set the working temperature of the process so as to oxidize all organic compounds. Both materials Pd/HY and CoAlCeO stand with the lowest P values of 7 and 20 °C respectively, sign of their improved performance for benzene oxidation.

To achieve real VOCs total oxidation, the working temperature must be set assuming the total oxidation of the reactant and its byproducts. This fact is supported by the toxic and regulatory aspects related to benzene. Therefore, the classification of the catalysts by their performance must be reassessed. Subsequently, according to the T_{50} and T_{100}, the ranking is as follows, from the most efficient to the least efficient: Pd/γ-Al$_2$O$_3$ > Pd/α-Al$_2$O$_3$ > Pd/CeO$_2$ > Pd/HY > CoAlCeO. Considering the emissions of benzene, as T_f, this classification may be reviewed as follows: Pd/γ-Al$_2$O$_3$ > Pd/HY > CoAlCeO > Pd/CeO$_2$ > Pd/α-Al$_2$O$_3$. The difference between the T_{100} value of Pd/γ-Al$_2$O$_3$ and Pd/HY catalysts is then reduced from 42 °C to 5 °C by considering the T_f value corresponding to the temperature at which toluene is totally converted. Thus, the performances of these two materials are very close. Moreover, if the CoAlCeO catalyst exhibits lower performance considering only T_{50} and T_{100} values, it is by far the lowest benzene emitter and consequently, accounting for this point, CoAlCeO is among the best catalysts for toluene oxidation.

2.3. Butanone Oxidation

After analyzing the efficiency of the different catalytic materials for toluene oxidation, these materials have been studied in the framework of butanone (MEK) oxidation with a similar methodology. The light-off curves obtained for all materials are shown in Figure 3.

The values of T_{50} and T_{100} for MEK oxidation are reported in Table 3 in comparison with values corresponding to the toluene oxidation. The test conducted with SiC shows a slight conversion from 260 °C, with a MEK conversion of 20% at 400 °C. As in the case of toluene, this observation can be explained by thermal decomposition. However, this phenomenon is more important because ketone function promotes the oxidation of the molecule. For the remaining materials, firstly, the curves show greater reactivity of MEK as compared to toluene. Indeed, besides the Pd/α-Al$_2$O$_3$ catalyst, the light-off curves are shifted to lower temperatures. This effect is particularly important for the CoAlCeO catalyst with a difference of 71 °C in the T_{50} values. Moreover, the MEK conversion at low temperature is particularly marked for the Pd/γ-Al$_2$O$_3$ and CoAlCeO catalysts. Therefore, from an activity point of view, the classification observed for the toluene oxidation on these materials is different, taking into account only the values of T_{100}. In the case of toluene, the performance sequence is: Pd/γ-Al$_2$O$_3$ > Pd/α-Al$_2$O$_3$ > Pd/CeO$_2$ > Pd/HY > CoAlCeO. For the MEK oxidation however, the performance sequence turns as follows: CoAlCeO > Pd/γ-Al$_2$O$_3$ > Pd/HY > Pd/CeO$_2$ > Pd/αAl$_2$O$_3$, where the CoAlCeO catalyst is now the most efficient catalyst.

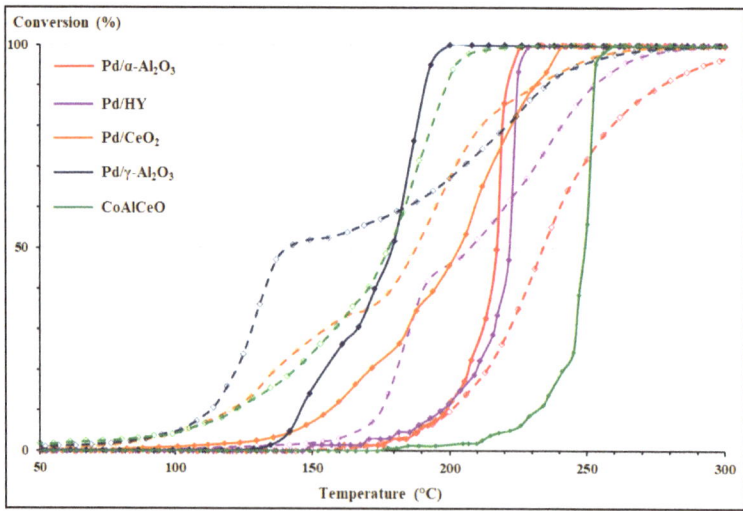

Figure 3. Comparison of light-off curves for toluene (full lines) and MEK (dashed lines) total oxidation.

Table 3. Comparison of catalyst performances for the MEK and toluene catalytic oxidation.

	MEK			Toluene		
	T_{50} (°C)	T_{100} (°C)	$T_{90}-T_{10}$ (°C)	T_{50} (°C)	T_{100} (°C)	$T_{90}-T_{10}$ (°C)
Pd/α-Al$_2$O$_3$	234	350	75	218	233	23
Pd/HY	207	312	77	222	242	28
Pd/CeO$_2$	189	347	111	197	235	75
Pd/γ-Al$_2$O$_3$	140	303	122	179	200	45
CoAlCeO	178	238	76	249	260	20

Figure 3 compares the light-off curves of toluene and MEK and clearly shows a shift of the light-off curves towards lower temperatures for MEK (dashed lines) relative to toluene (full lines). However, it is also possible to notice that these light-off curves are extended over a wider range of temperature in case of MEK oxidation. This is particularly highlighted by considering the differences between T_{90} and T_{10} values for each light-off curve (given in Table 3). In fact, this difference is on average 38 °C for toluene oxidation while it is on average 92 °C for MEK oxidation. For Pd/γ-Al$_2$O$_3$, Pd/CeO$_2$ and Pd/HY, the difference between the two light-off curves are substantially reduced and even reversed at high conversion: toluene indeed becomes easier to oxidize than MEK beyond a certain conversion. In addition, the light-off curves of MEK of these three catalysts exhibit an inflection point between 30% and 50% of conversion. From this point, light-off curves are stretched towards higher temperatures. In the case of Pd/α-Al$_2$O$_3$, no strong inflection point was evidenced, but the light-off curve stretches progressively with increasing temperatures. On the contrary, the light-off curve of the CoAlCeO material shows a similar behavior for both VOCs. The most important observation from this figure is the fact that the CoAlCeO catalyst achieves 100% conversion for the lowest temperature, with an important difference of approximately 65 °C with Pd/γ-Al$_2$O$_3$ and Pd/HY materials.

The profiles of the light-off curves for the MEK oxidation can be explained by the formation of byproducts. Indeed, the generation of byproducts in the case of oxygenated compounds' oxidation is relatively well known, and byproducts of MEK oxidation were identified as a result of various studies made on its partial oxidation and total oxidation [8,9,12,27,28]. Firstly, studies of McCullagh et al. [27,28], Álvarez-Galván et al. [12] and Arzamendi et al. [8] have identified three oxidation pathways that lead to several oxygenated compounds in C$_2$ and C$_4$: acetic acid, acetaldehyde, diacetyl (butane-2,3-dione) and methyl vinyl ketone. Secondly, a study realized by Machold et al. [9] has highlighted a fourth method of oxidation that leads to the formation of propanal, propanoic acid, and methanoic acid. The study of the byproducts has been applied for MEK oxidation. A microGC analysis permitted the detection of four peaks corresponding to oxidation by-products. Injection of organic standards allowed the identification of five compounds, in agreement with the literature: acetic acid (AcOOH), acetaldehyde (AcO), propionic acid (PrOOH), diacetyl (DAC), and methyl vinyl ketone (MVK). In fact, with acetic acid and acetaldehyde being co-eluted, it is not possible to differentiate and quantify them separately by micro-GC. Nevertheless, in contrast to acetic acid, acetaldehyde is carcinogenic and thus an unwanted product. Consequently, the corresponding chromatographic peak was assigned to acetaldehyde in order to consider the worst results in terms of byproduct emissions. The different byproducts have been quantified and their emission profiles have been determined as a function of MEK conversion and temperature, and are represented in Figure 4 for Pd/γ-Al$_2$O$_3$ (Figure 4a) and CoAlCeO (Figure 4b).

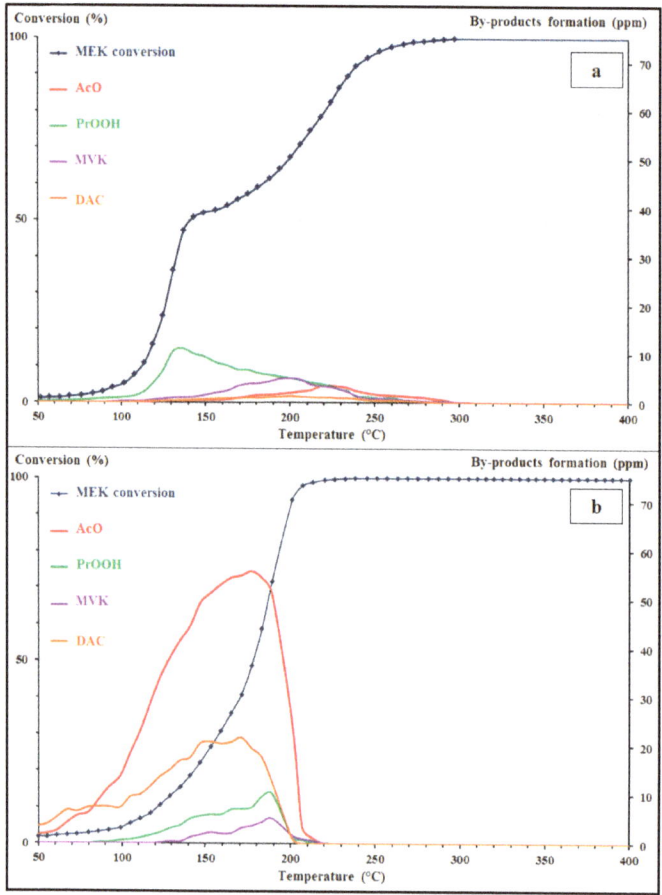

Figure 4. MEK conversion and MEK byproducts formation versus temperature for (**a**) Pd/γ-Al$_2$O$_3$ and (**b**) CoAlCeO catalysts (AcO: acetaldehyde; PrOOH: propionic acid; MVK: methyl-vinyl-ketone; DAC: diacetyl).

For the two materials Pd/γ-Al$_2$O$_3$ and CoAlCeO illustrate the two behaviors observed for the MEK oxidation. For Pd/γ-Al$_2$O$_3$, Figure 4a shows that the light-off curve is influenced by the generation of byproducts. In fact, the inflection point at 50% conversion is consistent with an emission peak of a byproduct: propionic acid. For the Pd/CeO$_2$ and Pd/HY materials, this inflection point is consistent with the emission peak of acetaldehyde. A hypothesis for this behavior is that these byproducts induce a partial inhibition of the MEK oxidation by blocking the catalytic sites. For CoAlCeO, the emission profiles are more important, especially for acetaldehyde (Figure 4b). However, despite this larger amount of byproducts, the material does not show any inhibitory effect on the MEK oxidation. The parameters used to characterize the benzene emission profiles were also applied on the byproducts' oxidation of MEK and are presented in Table 4.

Table 4. Parameters characterizing the emission profiles of MEK byproducts.

Pd/α-Al$_2$O$_3$	Q_{max} (ppm)	T_f (°C)	P (°C)	Pd/HY	Q_{max} (ppm)	T_f (°C)	P (°C)
AcO	5.4	362	+12	AcO	15.0	312	0
PrOOH	0.6	274	−76	PrOOH	11.4	312	0
MVK	2.9	350	0	MVK	1.3	312	0
DAC	2.8	322	−28	DAC	1.8	282	−30
Pd/CeO$_2$	Q_{max} (ppm)	T_f (°C)	P (°C)	**Pd/γ-Al$_2$O$_3$**	Q_{max} (ppm)	T_f (°C)	P (°C)
AcO	51.2	331	−16	AcO	3.6	309	+6
PrOOH	11.3	283	−64	PrOOH	11.1	303	0
MVK	3.4	300	−47	MVK	5.1	309	+6
DAC	5.6	244	−103	DAC	1.4	291	−12
CoAlCeO	Q_{max} (ppm)	T_f (°C)	P (°C)				
AcO	55.8	220	−18				
PrOOH	10.3	226	−12				
MVK	5.2	226	−12				
DAC	21.7	207	−31				

Table 4 reports the values of Q_{max}, T_f and P for MEK byproducts. The profiles show that acetaldehyde is the major byproduct of the MEK oxidation, except for Pd/γ-Al$_2$O$_3$. Moreover, it is possible to distinguish two emission profiles. The first can be observed for Pd/α-Al$_2$O$_3$, Pd/HY, and Pd/γ-Al$_2$O$_3$, where the maximum quantities issued for each byproduct are between 1 and 15 ppm. The second concerns Pd/CeO$_2$ and CoAlCeO, where these values remain on the same order of magnitude for propionic acid, methyl vinyl ketone and diacetyl, but acetaldehyde is emitted in a much larger quantity, up to 56 ppm. Finally, except for the two materials with alumina support, the byproducts seem to be fully oxidized before reaching the total conversion of MEK (T_{100}). For Pd/α-Al$_2$O$_3$ and Pd/γ-Al$_2$O$_3$, Table 4 shows a persistence of acetaldehyde. Nevertheless, this persistence is very low (6–12 °C) and has a small impact on the performance of these materials. Therefore, the sequence of performance established in relation to activities (T_{100}) remains the same after considering the byproducts: CoAlCeO > Pd/γ-Al$_2$O$_3$ > Pd/HY > Pd/CeO$_2$ > Pd/α-Al$_2$O$_3$.

2.4. Toluene/MEK Mixture Oxidation

The next step towards application of industrial exhaust is to study the behavior of the different catalytic materials for the total oxidation of a binary mixture (toluene/MEK 1000/1000 ppm). The results presented in this section are discussed according to two aspects: (i) the light-off curves of VOCs alone and in mixture are compared to assess mixing effects between toluene and butanone for each catalyst; (ii) the light-off curves of the binary mixture are compared in order to differentiate the performance of each catalyst. Two methods are used to calculate the VOC conversion. The first is shown in Equation (1) and denoted "$X_{Total\ Carbon}$". This allows for tracing the light-off curve in total carbon without taking into account the pressure drop and adsorption/desorption effects. Subsequently, the light-off curve is only influenced by chemical phenomena. However, this method gives a general idea and does not allow following the conversion of each VOC in the mixture. The second method is a conversion calculation taking into account the initial and outgoing quantity of VOCs. However, the light-off curve is influenced by physical phenomena (pressure drop, adsorption/desorption phenomena). The formula is given by Equation (2) and is denoted "X_{VOC}". Thus, the first method has been used to compare oxidation performances between catalysts, and the second to compare the oxidation performances between VOCs alone and in mixture.

$$X_{Total\ Carbon} = 100 \times \frac{\sum X_i \times P_{i,T} + CO_{2,T}}{\sum X_i \times P_{i,T} + CO_{2,T} + \sum X_i \times R_{i,T}} \quad (1)$$

$$X_{VOC} = 100 \times \frac{R_{i,0} - R_{i,T}}{R_{i,0}}, \qquad (2)$$

where:

$R_{i,0}$ is the initial mole percentage of VOCs;

$R_{i,T}$ is the mole percentage of VOCs at the temperature T;

$P_{i,T}$ is the mole percentage of byproducts at the temperature T;

$CO_{2,T}$ is the mole percentage of carbon dioxide at the temperature T;

X_i is the number of carbon atoms in the compound i.

Concerning the mixing effects, two behaviors have been identified: inhibitory effects as well as beneficial effects. These are more or less marked as a function of the materials used and the temperature. As for the influence of MEK byproducts, these effects are more important on Pd/γ-Al$_2$O$_3$ and CoAlCeO catalysts. The light-off curves of the binary mixture for these catalysts are shown in Figure 5, and the corresponding T_{50} values are reported in Table 5.

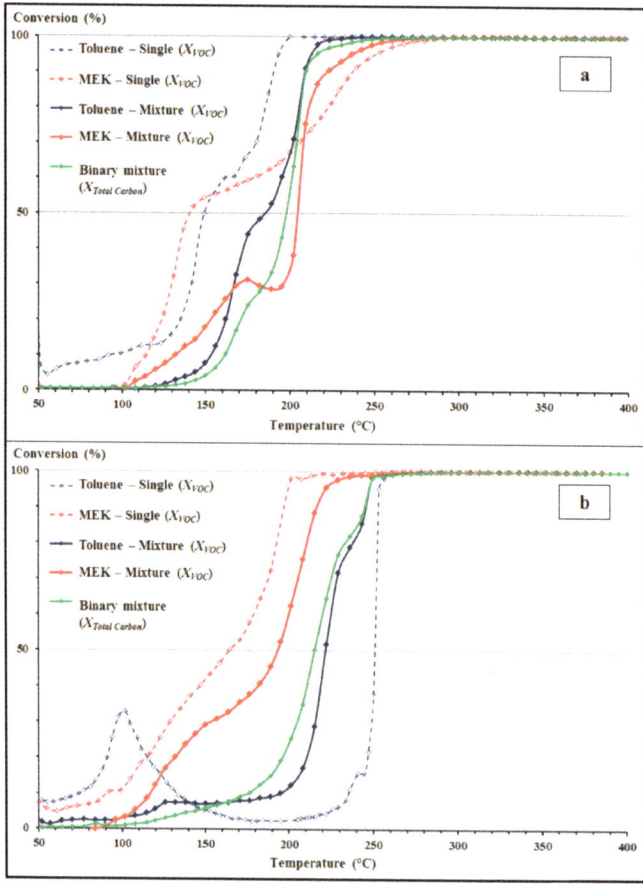

Figure 5. Light-off curves of VOCs alone and in mixture (X_{VOC} formula) and binary mixture ($X_{Total\ Carbon}$ formula) for (**a**) Pd/γ-Al$_2$O$_3$ and (**b**) CoAlCeO catalysts.

Table 5. Comparison of T_{50} for the total catalytic oxidation of binary mixture for Pd/γ-Al$_2$O$_3$ and CoAlCeO catalysts.

	Pd/γ-Al$_2$O$_3$		CoAlCeO	
	Toluene (°C)	MEK (°C)	Toluene (°C)	MEK (°C)
VOC alone	149	141	251	165
VOC in mixture	185	204	221	193

In the case of Pd/γ-Al$_2$O$_3$ catalyst (Figure 5a), a modification of the reactivity is revealed for both VOCs. For MEK oxidation in mixture, a shift of the light-off curve to higher temperatures is observed as compared to the light-off curve of MEK alone. Indeed, the shift between the curves begins at 100 °C on a conversion range of 0 to 70%, with a value of 63 °C at T_{50}. Beyond 70% of conversion, both light-off curves are reversed, MEK being then slightly more reactive in the presence of toluene. For toluene oxidation in mixture, a shift in the conversion curve is also observed with the light-off curve of toluene alone. Consequently, the light-off curves of VOCs in a binary mixture show inhibitory effects for both compounds. This can be explained by competitive reactions between the two molecules in the adsorption/desorption steps, causing the blocking of the active sites of the catalyst. This phenomenon has already been observed in the literature in mixtures of hydrocarbons and oxygenated compounds [7,13,29,30]. These inhibitory phenomena are also observed for Pd/HY and Pd/CeO$_2$.

For the CoAlCeO material (Figure 5b), the results also show a change in the reactivity of both VOCs. For MEK oxidation in mixture, a shift of the light-off curve to higher temperatures is observed. However, this effect between the two curves is more regular than with Pd/γ-Al$_2$O$_3$. A shift of 28 °C at T_{50} with the light-off curve of MEK alone is observed. In contrast, a beneficial effect is observed on the toluene oxidation in mixture. Indeed, the light-off curve of toluene in mixture is clearly shifted to lower temperatures (a difference of 30 °C at T_{50} with the toluene-alone light-off curve). This behavior can be explained by an additive effect, as has been highlighted in the studies of Beauchet et al. [31,32] for the oxidation of xylene/isopropanol mixtures. It has been shown that an intermediate compound was formed during the oxidation of this mixture, isopropyldimethylbenzene, which is obtained from the alkylation of xylene with propene, an oxidation byproduct of isopropanol. This aromatic intermediate compound possesses more alkyl chains than xylene and a ternary carbon atom, therefore its reactivity is probably higher than that of xylene. Hence, the formation of this intermediate compound lowers the temperature of xylene oxidation due to the indirect increase of its reactivity. Is has to be noted that in the case of xylene and isopropanol, the number of their respective byproducts is restricted [31,33], leading to a facilitated identification of these intermediates during the oxidation of xylene/isopropanol mixtures. In this work, both VOCs have a greater number of byproducts with almost similar structures and chemical properties [3,8,9,33]: the identification of a potential byproduct from the addition of toluene and MEK is then more difficult. Additional analyses were performed with a Omnistar Quadrupole Mass Spectrometer (QMS-200) (Pfeiffer Vacuum, Asslar, Germany), configured in "Bargraph" mode to scan all m/z fragments between 1 and 199. These analyses have failed to detect compounds other than those already known and mentioned. As a consequence, if an additional compound is formed between the toluene and MEK, its lifetime is probably too short to be detected by ex situ analysis. Enhanced catalytic activity has also been observed on the Pd/α-Al$_2$O$_3$ catalyst, but the difference was not significant.

Concerning the oxidation of the binary mixture, the light-off curves obtained are shown in Figure 6 for all materials. The values of T_{50} and T_{100} are reported in Table 6, as well as the values for the oxidation of toluene for comparison.

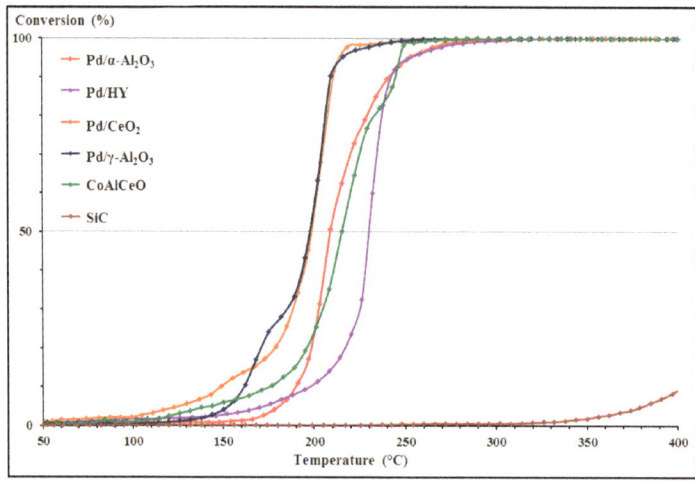

Figure 6. Light-off curves of binary mixture total oxidation.

Table 6. T_{50} and T_{100} of catalysts for binary mixture and toluene catalytic oxidation.

Catalyst	Binary Mixture		Toluene	
	T_{50} (°C)	T_{100} (°C)	T_{50} (°C)	T_{100} (°C)
Pd/α-Al$_2$O$_3$	209	326	218	233
Pd/HY	230	330	222	242
Pd/CeO$_2$	199	314	197	235
Pd/γ-Al$_2$O$_3$	198	290	179	200
CoAlCeO	215	290	249	260

The test conducted with SiC shows a slight conversion of the binary mixture from 225 °C, with a maximum of 9.9% at 400 °C. This partial conversion is derived from the thermal decomposition of VOCs, primarily of MEK, as has been observed for blank experiments performed with pure VOCs. In addition, partial conversion of the mixture begins at 225 °C instead of 260 °C and 360 °C, respectively, for MEK and toluene used on their own. However, partial conversion of the mixture reached a maximum of 9.9% (198 ppm) at 400 °C instead of 20.6% (206 ppm) and 5.7% (57 ppm), respectively, for MEK and toluene. Thus, the binary mixture seems to be reactive at lower temperatures, but with more limited conversion rates at high temperatures. As a first proof, the tests show light-off curves much less stretched than in the case of MEK oxidation. In addition, mixing effects observed for palladium-based catalysts do not significantly modify their performance. Indeed, given the T_{50} values, they remain close to those observed during toluene oxidation, with only a small temperature difference. Nevertheless, the mixing effects observed for the CoAlCeO mixed oxide have a strong impact on its activity: this is the least active catalyst for oxidation of toluene alone, without considering the byproduct oxidation. For the binary mixture, the presence of the MEK significantly lowers the value of T_{50} and this material exhibits performance fully comparable to palladium-based catalysts. As a second proof, beyond 95% of conversion, the light-off curves significantly stretch, which is correlated with a strong persistence of the residual traces of VOCs in the flow, especially for MEK. Therefore, the T_{100} values for the binary mixture oxidation are more important for all materials with respect to the oxidation of toluene alone, with an average shift of 88 °C. This phenomenon was expected since the total VOC concentration is higher in the binary mixture (2000 ppm instead of 1000 ppm). An important point is that only CoAlCeO mixed oxide presents a reduced impact of this phenomenon with only 30 °C difference on its T_{100} value. As a conclusion, this oxide corresponds to one of the best catalyst for the total abatement of the binary mixture. According to the T_{50} and T_{100}, the performance sequence observed for the binary mixture oxidation is the following: CoAlCeO = Pd/γ-Al$_2$O$_3$ > Pd/CeO$_2$ > Pd/α-Al$_2$O$_3$ > Pd/HY.

Regarding the effect of byproducts on the behavior of the mixture, their profiles were followed according to the conversion and temperature. The results show the emission of all byproducts already detected during the study of toluene and MEK oxidation: benzene, acetaldehyde (AcO), propionic acid (PrOOH), methyl vinyl ketone (MVK), and diacetyl (DAC). Additionally, the emission profiles present significant changes due to the mixing effects. Indeed, the persistence, P, and the maximum quantity emitted observed Q_{max}, are significantly changed. Figure 7 shows the emission profiles of the byproducts from the binary mixture oxidation on Pd/CeO$_2$.

This catalyst exhibits the most remarkable and representative impacts of mixing effects on byproducts' emission profiles. Table 7 reports values of Q_{max}, T_f, and P for the binary mixture, as well as the values for oxidation of toluene and MEK alone for comparison.

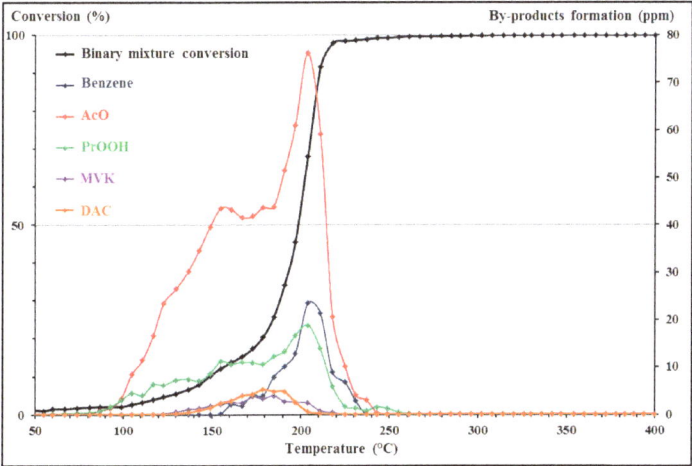

Figure 7. Binary mixture conversion and byproduct formation versus temperature for Pd/CeO$_2$ catalyst.

Table 7. Parameters characterizing the emission profiles on Pd/CeO$_2$ of byproducts for VOCs alone and in binary mixture.

Toluene	Q_{max} (ppm)	T_f (°C)	P (°C)
Benzene	7.0	298	63
MEK	Q_{max} (ppm)	T_f (°C)	P (°C)
AcO	51.2	331	−16
PrOOH	11.3	283	−64
MVK	3.4	300	−47
DAC	5.6	244	−103
Binary Mixture	Q_{max} (ppm)	T_f (°C)	P (°C)
Benzene	23.6	237	−77
AcO	76.1	249	−65
PrOOH	18.8	267	−47
MVK	4.0	225	−89
DAC	5.2	218	−96

The mixing effects mainly impact on the values of P and Q_{max}. Regarding the persistence, P, the values decrease significantly for several compounds. This observation indicates that byproducts are eliminated at a lower temperature. This is particularly true for benzene whose P value is +63 °C for toluene oxidation instead of −77 °C for binary mixture oxidation. As mentioned before, this effect is especially beneficial due to the toxicity of this compound. Propionic acid is the only byproduct here whose persistence increases significantly, although the value is still negative. Concerning Q_{max} values, the emission peaks of benzene, acetaldehyde, and propionic acid are observed. This proves that even though these compounds are more easily oxidized, they are also emitted in larger amounts. In a different way to these three compounds, the values of Q_{max} observed for the methyl vinyl ketone and diacetyl show very slight variations, and these two compounds are present as minority byproducts. This observation is also carried out for all materials studied on the binary mixture. To simplify the comparison between materials, only benzene, acetaldehyde, and propionic acid will be considered thereafter. Table 8 reports the Q_{max}, T_f, and P values of these three compounds for the five materials. This table highlights the significant reduction in the persistence of benzene.

Table 8. Parameters characterizing the emission profiles of binary mixture byproducts.

Pd/α-Al$_2$O$_3$	Q_{max} (ppm)	T_f (°C)	P (°C)	Pd/HY	Q_{max} (ppm)	T_f (°C)	P (°C)
Benzene	11.5	361	+35	Benzene	9.9	282	−48
AcO	7.6	320	−6	AcO	8.5	336	+6
PrOOH	1.3	314	−12	PrOOH	4.8	306	−24
Pd/CeO$_2$	Q_{max} (ppm)	T_f (°C)	P (°C)	Pd/γ-Al$_2$O$_3$	Q_{max} (ppm)	T_f (°C)	P (°C)
Benzene	23.6	237	−77	Benzene	26.5	296	+6
AcO	76.1	249	−65	AcO	3.8	266	−24
PrOOH	18.8	267	−47	PrOOH	11.6	260	−30
CoAlCeO	Q_{max} (ppm)	T_f (°C)	P (°C)				
Benzene	4.0	272	−18				
AcO	73.7	255	−35				
PrOOH	15.5	236	−54				

Indeed, except for the two materials with an alumina support, the persistence of benzene is negative, but has nonetheless been greatly reduced since these values have increased from +6 °C to +35 °C and +44 °C instead of +67 °C, respectively, for the oxidation of toluene alone. Regarding acetaldehyde and propionic acid, the tests show a relative increase in Q_{max} values as compared to those for MEK alone. Similarly, the persistence of acetaldehyde emissions is greatly reduced for all materials except Pd/HY. For the persistence of propionic acid, the behavior is more variable, though: a slight increase is observed for Pd/α-Al$_2$O$_3$ and Pd/CeO$_2$, while a decrease is observed for Pd/HY, Pd/γ-Al$_2$O$_3$ and CoAlCeO. Considering these facts, the study of the total oxidation of the binary mixture highlights the two most efficient catalytic materials: the commercial formulation Pd/γ-Al$_2$O$_3$ and the mixed oxide CoAlCeO. Indeed, these materials show the same performance in terms of activity with a T_{100} to 290 °C for both materials. Considering the byproducts, the mixed oxide has a slight advantage over Pd/γ-Al$_2$O$_3$ since the byproducts are completely eliminated at 290 °C. Considering this fact, the performance sequence is as follows: CoAlCeO > Pd/γ-Al$_2$O$_3$ > Pd/CeO$_2$ > Pd/HY > Pd/αAl$_2$O$_3$.

2.5. Oxidation of a Simulated Industrial Exhaust

The literature on VOCs' oxidation systematically studied models exhausts where VOCs' concentration and flow rate are perfectly controlled, while only a few articles focus on the study of VOCs mixtures [7,13–18,31,32]. As in this study, these articles highlight different inhibitory or beneficial effects derived from interactions between compounds. However, a binary or ternary mixture studied at the laboratory level remains far from the reality of an industrial effluent, which can be made up of many more compounds (10 to 50 compounds) with variable compositions and concentrations over time. To get closer to industrial applications, we further studied a more complex mixture of VOCs that was established based on the composition of an industrial effluent and will therefore be referred to as a simulated industrial exhaust. In the industry, part of the manufacturing process is to apply lacquers and paints to metal surfaces, the consequence being that industry emits significant amounts of gaseous effluents with a high concentration of VOCs and a high flow rate. These effluents contain mostly aromatic compounds, mainly toluene, and oxygenated compounds, mainly MEK, as well as traces of paraffins (C_9–C_{11}). The liquid mixture used here to simulate the industrial effluent of VOCs was constituted by selecting six aromatic compounds (toluene, o-xylene, ethylbenzene, propylbenzene, 4-ethylbenzene, and 1,4diethylbenzene), three oxygenates (butanone, butanol, butoxyethanol), and a paraffin (decane). The concentration of each substance is fixed to a concentration of 10 vol% in the liquid phase. This mixture was studied on the same experimental setup as in previous studies, the mixture being placed in a saturator. The temperature of the saturator was set so as to fix a value of 7000 ppm carbon equivalent (ppm eq. C) in a 100 mL·min^{-1} air flow, with a total hydrocarbons analyzer COSMA Graphite 655 (Environnement SA, Poissy, France). The gas phase composition determined by microGC is given in Figure 8.

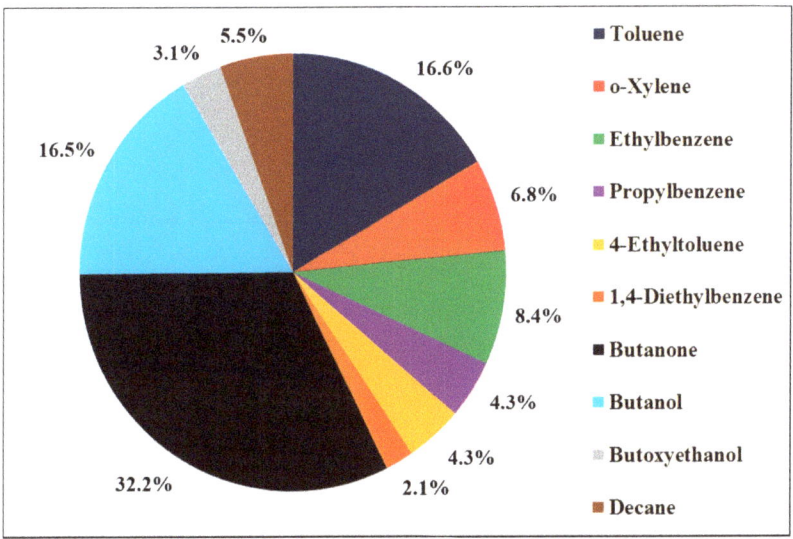

Figure 8. Composition of the complex mixture determined by μGC.

In order to place themselves under real measurement conditions, the abatement rate of VOCs mixture is measured with a total hydrocarbons analyzer (COSMA Graphite 655), equivalent to a flame ionization detector (FID). Indeed, VOC measurements on industrial effluents are performed with this type of device, and this measuring method conforms to a French regulatory norm (NF X 43-301). Furthermore, although the microGC is a suitable tool for measuring VOCs' oxidation, this technique reaches its limits with regards to the detection and separation of the various molecules that can be emitted or generated during the process. The abatement curves obtained by the total hydrocarbons analyser (measurements given in equivalent carbon (eq. C)) are comparable to light-off curves determined by microGC. Nevertheless, microGC was used as a complementary analytic device, as well as trapping on cartridge Tenax TA (Supelco, Bellefonte, PA, USA) and Mass Spectrometer QMS-200 (Pfeiffer Vacuum, Asslar, Germany), for speciation analysis. Finally, this study has been conducted with the two catalysts presenting the best performances: Pd/γ-Al_2O_3 and CoAlCeO. These two materials appear to be most effective for the abatement of the VOCs studied and their resulting byproducts. This choice is relevant for two main reasons: Firstly, Pd/γ-Al_2O_3 is an efficient commercial catalyst that is sold for such industrial applications and is thus suitable as a benchmark. Secondly, CoAlCeO is an alternative formulation exhibiting an efficiency similar to or even better than the commercial material. In addition, this material is not made of noble metals and its cost is considerably lower as compared to the commercial formulation. The abatement curves obtained, expressed in ppm carbon equivalent (ppm eq. C) versus temperature, are given in Figure 9. The T_{50} values are reported in Table 9, where the values for the oxidation of toluene and binary mixture are also reported for comparison.

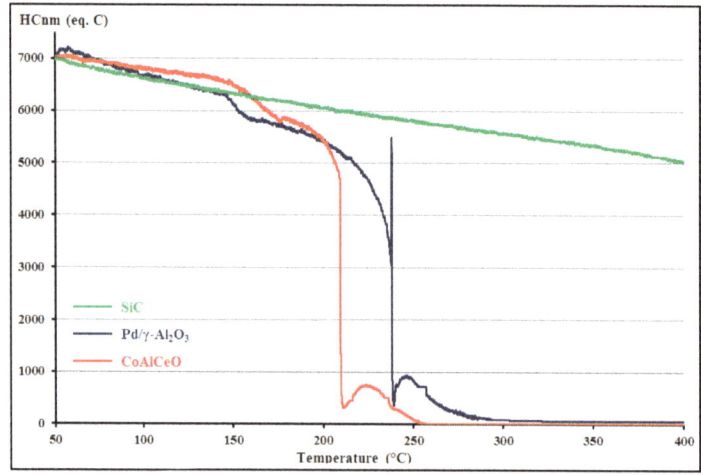

Figure 9. Abatement curves of the complex mixture on Pd/γ-Al$_2$O$_3$ and CoAlCeO catalysts.

Table 9. T_{50} of Pd/γ-Al$_2$O$_3$ and CoAlCeO catalysts for toluene, binary mixture, and complex mixture total oxidation.

	Toluene (°C)	Binary Mixture (°C)	Complex Mixture (°C)
Pd/γ-Al$_2$O$_3$	179	198	237
CoAlCeO	249	215	210
Δ	+70	+17	−27

The test performed on the SiC shows a partial conversion of the complex mixture, with a maximum of 28.1% at 400 °C. This relatively high conversion is due to the thermal decomposition of oxygenates (MEK, butan-1-ol and butoxyethanol). This was confirmed by microGC analyses that highlight some byproducts: acetaldehyde, propionic acid, methyl vinyl ketone and diacetyl, which are derived from the MEK oxidation and butanal and 1-butene, which are derived from the butan-1-ol conversion (partial oxidation and dehydration, respectively). In addition, analyses do not show the presence of CO_2, which confirms partial oxidation only. The results show an abatement of 99.2% and 100% of the complex mixture for Pd/γ-Al$_2$O$_3$ and CoAlCeO, respectively. In addition, the abatement curves do not exhibit a sigmoidal profile characteristic of the light-off curves. The conversion observed between 0% and 20% is due to the decomposition of oxygenates, mainly butan-1-ol and butoxyethanol, leading to the corresponding aldehydes and acids (butanal, acetaldehyde, and acetic acid). The degradation of aromatic compounds is only starting at 155 and 175 °C for CoAlCeO and Pd/γ-Al$_2$O$_3$, respectively. Then, the mixture conversion abruptly increases from 30 to 95% for CoAlCeO and from 55 to 95% for Pd/γ-Al$_2$O$_3$ in just seconds. This is accompanied by an overpressure/depression phenomenon, confirmed by mass spectrometry, and an immediate increase of 30 °C of the reactor temperature. This is characteristic of a deflagration effect, which explains the appearance of abatement curves.

This phenomenon is slower on Pd/γ-Al$_2$O$_3$, and also more visible. Moreover, a peak of methane emission is measured by the total hydrocarbons analyzer. Finally, the system returns to a steady state, with a slight loss of conversion, before the degradation of the residual organic fraction is performed until the maximum conversion is reached.

Concerning the performance, Pd/γ-Al$_2$O$_3$ presents a T_{50} values around 40 °C greater than in the binary mixture case, highlighting a more significant inhibition effect. In addition, conversion of the complex mixture stabilizes at 99.2% at 325 °C. Additional analyses at 325 (T_{99}), 350, and 400 °C show that the remaining organic fraction is predominantly composed of decane and benzene, as well as a small proportion of MEK. For CoAlCeO, the T_{50} value is lowered once more in comparison with the binary mixture and toluene oxidation, but to a lower extent with only 5 °C on T_{50} between the complex and binary mixture. Moreover, conversion of the complex mixture reaches 100% from 357 °C. Additional analysis at 325 (T_{99}) and 350 °C show that the residual organic fraction is composed of MEK and toluene, with MEK being predominant. The same analysis conducted at 400 °C confirms a total abatement of the mixture. Therefore, the CoAlCeO mixed oxide is highlighted as a promising alternative material to the commercial Pd/γ-Al$_2$O$_3$ catalyst, with the latter currently being used for industrial units' VOCs treatment. Indeed, CoAlCeO shows the best performance for the complex mixture oxidation considering the rate and temperature of VOCs abatement. In addition, no toxic compounds seem to be emitted at high conversion by this catalyst.

3. Materials and Methods

3.1. Synthesis of the Palladium Impregnated Materials

3.1.1. Preparation of Supports

Ceria support was prepared by a precipitation method. A solution of Ce(NO$_3$)$_3$·6H$_2$O (Fisher Scientific, Hampton, NH, USA) was added drop by drop to a NaOH (Fisher Scientific, Hampton, NH, USA) solution with a molar ratio Ce^{3+}/OH$^-$ of 1/5. The addition was made under magnetic stirring over 3 h. The suspension was left under stirring for 2 h at ambient temperature. Then, the suspension was filtered and washed six times with 200 mL of hot deionized water (~60 °C). The solid was dried for 24 h at 100 °C and calcined for 4 h at 500 °C (1 °C·min^{-1}) under air flow (2 L·h^{-1}).

HY Faujasite zeolite was prepared by ionic exchange. A commercial NaY Faujasite zeolite (Si/Al: 2.7; Sigma-Aldrich, St. Louis, MO, USA) was slurried in a solution of ammonium nitrate 2.0 mol·L^{-1} with a molar ration NH$_4^+$/Na$^+$ of 20. The suspension was maintained for 18 h at 80 °C under magnetic agitation. Then, the suspension was filtered and washed with hot deionized water (~60 °C) to a neutral pH. These operations were repeated three times. Then, the solid was dried overnight at 100 °C and calcined for 4 h at 400 °C (1 °C·min^{-1}) under air flow (2 L·h^{-1}).

For the α-Al$_2$O$_3$, a commercial powder was chosen (ACROS ORGANICS, 99.0%).

3.1.2. Palladium Impregnation

Palladium (Fisher Scientific, Hampton, NH, USA) was deposed on support (α-Al$_2$O$_3$, CeO$_2$, and HY Faujasite) by aqueous impregnation method (0.5 wt % Pd). Solid was suspended in an appropriate volume of palladium nitrate solution (0.25 g·L^{-1}). The suspension was maintained for 18 h at 60 °C. Then, water was removed by a rotary evaporator. The solid was dried overnight at 100 °C and then calcined for 4 h at 400 °C (1 °C·min^{-1}) under air flow (2 L·h^{-1}).

A commercial Pd/γ-Al$_2$O$_3$ catalyst (ACROS ORGANICS, 0.5 wt %) was chosen to complete the study.

3.2. Synthesis of the Mixed Oxide

The CoAlCe mixed oxide was synthesized in the hydrotalcite way. An aqueous solution of 200 mL was prepared with Co(NO$_3$)$_2$·6H$_2$O, Al(NO$_3$)$_3$·9H$_2$O and Ce(NO$_3$)$_3$·6H$_2$O (Fisher Scientific, Hampton,

NH, USA), with a molar ratio of $Co^{2+}/Al^{3+}/Ce^{3+}$ of 6/1.2/0.8 ($Co_6Al_{1.2}Ce_{0.8}O_x$). This solution was added drop by drop to 30 mL of a Na_2CO_3 (Fisher Scientific, Hampton, NH, USA) solution (1 mol·L^{-1}). The pH of the additive solution was maintained at a value of 10.5 with a NaOH solution (2 mol·L^{-1}). After addition, the suspension was stirred for 18 h at room temperature. Then, the latter was filtered and washed with hot deionized water (~60 °C). The solid obtained was dried in an oven for 24 h at 60 °C and ground before being calcined for 4 h at 500 °C (1 °C·min^{-1}) under air flow (2 L·h^{-1}) and given the code name CoAlCeO.

3.3. Characterization of the Catalysts

Crystalline structures were determined at room temperature from X-ray Diffraction (XRD) recorded on a D8 Advance diffractometer (Bruker AXS, Billerica, MA, USA) equipped with a copper anode (λ = 1.5406 Å) and a LynxEye Detector. The scattering intensities were measured over an angular range of $10° \leq 2\theta \leq 80°$ for all samples with a step size of $\Delta(2\theta)$ = 0.02° and a count time of 4 s per step. The diffraction patterns were indexed by comparison with the "Joint Committee on Powder Diffraction Standards" (JCPDS) files.

Specific surface areas (SBET) were determined by the Brunauer–Emmet–Teller method on N_2 isotherms measured with a ThermoElectron Qsurf M1 series Surface Area Analyzer apparatus (Waltham, MA, USA). The sample was degassed before measurement at 130 °C under helium flow. Adsorption was made with a 30% N_2–70% He mixture at −196 °C. Desorption of gaseous N_2 was quantified with a thermal conductivity detector.

For the elemental analysis, 50.0 mg of powder was dissolved in 5 mL of aqua regia (HNO_3/HCl 1:2) (Fisher Scientific, Hampton, NH, USA) under microwave for 30 min (CEM, Model MARSXpress, Matthews, CA, USA). Then, the solution was extended to 50.0 mL with ultrapure water and filtered with a 0.45-µm cellulosic micro-filter. Analysis was performed with an ICP-OES (Thermo, model ICAP 6300 Duo, Waltham, MA, USA).

Hydrogen chemisorption was realized with 0.5 g of catalysts mixed with 0.5 g of silicon carbide, in order to have a homogeneous catalytic bed and avoid hot spots. The mixture was reduced for 2 h at 200 °C (3 °C·min^{-1}) under dihydrogen flow (30 mL·min^{-1}). Then, the system was cooled down to 80 °C under an argon flow (30 mL·min^{-1}). A step of 30 min under argon flow (30 mL·min^{-1}) was realized, followed by a step of 30 min under 10% H_2/Ar flow (30 mL·min^{-1}) and a last step of 30 min under argon flow (30 mL·min^{-1}). Finally, desorption was performed under a flow of argon (30 mL·min^{-1}) from 80 to 475 °C (10 °C·min^{-1}), followed by a step of 10 min. The detector used was a mass spectrometer Balzers QMG 422 MS (Pfeiffer Vacuum, Asslar, Germany).

3.4. Catalytic Tests

Total oxidation of VOCs was studied in a fixed bed reactor loaded with 100 mg of catalyst. VOC/Air mixtures were generated with a saturator in order to obtain 1000 ppm of VOC in a flow of 100 mL·min^{-1}. A toluene/MEK binary mixture was generated in order to obtain 1000 ppm of each VOC in a flow of 100 mL·min^{-1}. Tests were made between 50 and 400 °C with a temperature ramp of 1.5 °C·min^{-1}. Catalysts were pre-treated for 2 h at 200 °C (1 °C·min^{-1}) under air flow (2 L·h^{-1}) and then, for impregnated catalysts, were reduced for 2 h at 200 °C (1 °C·min^{-1}) under dihydrogen (5.0) flow (2 L·h^{-1}). Organic compounds were analyzed by gas chromatography with a CP-4900 microGC (Agilent Technologies Inc., Santa Clara, CA, USA). A test was systematically conducted under the same conditions with silicon carbide SiC to achieve a blank experiment.

Catalytic performances were compared considering the T_{50} and T_{100}, which correspond to the temperature at which 50% and 100% of VOC, respectively, were converted. VOC conversion was calculated considering products and byproducts and as a function of the number of carbons for each compound:

$$X_T = 100 \times \frac{\sum X_i \times P_{i,T} + CO_{2,T}}{\sum X_i \times P_{i,T} + CO_{2,T} + \sum X_i \times R_{i,T}} \qquad (3)$$

where:

$P_{i,T}$ is the mole percentage of VOCs at the temperature T;
$P_{i,T}$ is the mole percentage of byproducts at the temperature T;
$CO_{2,T}$ is the mole percentage of carbon dioxide at the temperature T;
X_i is the carbon number of corresponding compounds.

Catalytic activity was calculated at a VOC conversion of 20% and considering a plug-flow reactor:

$$A = \frac{Q}{V_M} \cdot \frac{273.15}{T_{20}} \cdot \frac{R_{i,o}}{10^6} \cdot \frac{X}{m} \cdot \frac{1}{S_{BET}} \tag{4}$$

where:

Q is the volume flow (L·h^{-1});
V_M is the molar volume (L·mol^{-1});
T_{20} is the catalyst temperature for 20% VOC conversion (K);
$R_{i,o}$ is the VOC initial concentration (ppm);
X is the VOC conversion (%);
m is the catalyst mass (g);
S_{BET} is the specific surface area of the catalyst (m^2·g^{-1}).

4. Conclusions

Catalytic oxidation appears to be an efficient VOC treatment process that is more cost-effective and environmentally friendly than thermal oxidation. Nevertheless, the cost of the catalytic materials remains an obstacle to the wider use of this process at an industrial level. Indeed, the most efficient catalysts are generally constituted of precious metals (Pt, Pd, Pt-Pd). The aim of this work was to propose an alternative material to the materials commonly used for this application. To bring this to fruition, a CoAlCeO mixed oxide was developed in the laboratory, and was compared to palladium-based catalysts (Pd/α-Al$_2$O$_3$, Pd/CeO$_2$, Pd/HY and a commercial Pd/γ-Al$_2$O$_3$) for the total oxidation of industrial VOCs: toluene and butanone (MEK). These VOCs were selected as model molecules in agreement with their use in industry as a solvent in paints, inks and varnishes for application on metal surfaces. The study was conducted for these five materials on the total oxidation of toluene, MEK, and a toluene/MEK binary mixture. The study was also carried out taking into account the formation of byproducts in the catalytic performance. Tests have shown various results for the three sets of experiments presented in this study. Different performances, as well as inhibition or beneficial effects, were identified. Nevertheless, all the results highlight the two most effective materials for the oxidation of VOCs, but also for oxidation of their byproducts: Pd/γ-Al$_2$O$_3$ and CoAlCeO. The results obtained with the commercial formulation Pd/γ-Al$_2$O$_3$ were expected since it is one of the formulations typically used for this industrial process; its effectiveness has been already demonstrated. Therefore, our results on CoAlCeO mixed oxide highlight their efficiency and put forward this material as a relevant alternative to conventional Pd-based catalysts. The estimated cost of the CoAlCeO catalyst used in this study, based on the price of the metallic precursors and the chemical species used, is around 30% lower than the Pd/γ-Al$_2$O$_3$ catalyst. Thus, this oxide opens up the development of new, efficient and less expensive catalysts for this treatment process.

Acknowledgments: The authors wish to acknowledge the "Agency of Environment and Energy Management" (ADEME) as well as the Nord-Pas-de-Calais region for the funding of the work (project number 1281C0095). The authors also acknowledge the "Centre Commun de Mesures" of ULCO. Cédric Barroo thanks the Fonds de la Recherche Scientifique (F.R.S.-FNRS) for financial support.

Author Contributions: Julien Brunet and Eric Genty performed the material preparation and the experiments. Cédric Barroo, Fabrice Cazier, Christophe Poupin, Stéphane Siffert, Diane Thomas, Guy De Weireld, and Thierry Visart de Bocarmé revised and modified the paper. All authors contributed equally to the data interpretation and discussion. Julien Brunet, Eric Genty, Cédric Barroo, and Renaud Cousin wrote the final manuscript. Renaud Cousin conceived and managed the project.

Conflicts of Interest: The authors declare no conflict of interest.

References

1. Hu, C. Catalytic combustion kinetics of acetone and toluene over $Cu_{0.13}Ce_{0.87}O_y$ catalyst. *Chem. Eng. J.* **2011**, *168*, 1185–1192. [CrossRef]
2. Liotta, L.F.; Ousmane, M.; Di Carlo, G.; Pantaleo, G.; Deganello, G.; Boreave, A.; Giroir-Fendler, A. Catalytic removal of toluene over Co_3O_4–CeO_2 mixed oxide catalysts: Comparison with Pt/Al_2O_3. *Catal. Lett.* **2008**, *127*, 270–276. [CrossRef]
3. Brunet, J.; Genty, E.; Landkocz, Y.; Al Zallouha, M.; Billet, S.; Courcot, D.; Siffert, S.; Thomas, D.; De Weireld, G.; Cousin, R. Identification of by-products issued from the catalytic oxidation of toluene by chemical and biological methods. *Comptes Rendus Chim.* **2015**, *18*, 1084–1093. [CrossRef]
4. Genty, E.; Cousin, R.; Capelle, S.; Gennequin, C.; Siffert, S. Catalytic oxidation of toluene and CO over nanocatalysts derived from hydrotalcite-like compounds ($X_6^{2+}Al_2^{3+}$): Effect of the bivalent cation. *Eur. J. Inorg. Chem.* **2012**, *2012*, 2802–2811. [CrossRef]
5. Santos, V.P.; Pereira, M.F.R.; Órfão, J.J.M.; Figueiredo, J.L. Mixture effects during the oxidation of toluene, ethyl acetate and ethanol over a cryptomelane catalyst. *J. Hazard. Mater.* **2011**, *185*, 1236–1240. [CrossRef] [PubMed]
6. Wu, H.; Wang, L.; Zhang, J.; Shen, Z.; Zhao, J. Catalytic oxidation of benzene, toluene and *p*-xylene over colloidal gold supported on zinc oxide catalyst. *Catal. Commun.* **2011**, *12*, 859–865. [CrossRef]
7. Burgos, N.; Paulis, M.; Mirari Antxustegi, M.; Montes, M. Deep oxidation of VOC mixtures with platinum supported on Al_2O_3/Al monoliths. *Appl. Catal. B Environ.* **2002**, *38*, 251–258. [CrossRef]
8. Arzamendi, G.; de la Peña O'Shea, V.A.; Álvarez-Galván, M.C.; Fierro, J.L.G.; Arias, P.L.; Gandía, L.M. Kinetics and selectivity of methyl-ethyl-ketone combustion in air over alumina-supported PdO_x-MnO_x catalysts. *J. Catal.* **2009**, *261*, 50–59. [CrossRef]
9. Machold, T.; Suprun, W.Y.; Papp, H. Characterization of VO_x-TiO_2 catalysts and their activity in the partial oxidation of methyl ethyl ketone. *J. Mol. Catal. A Chem.* **2008**, *280*, 122–130. [CrossRef]
10. Gandía, L.M.; Gil, A.; Korili, S.A. Effects of various alkali-acid additives on the activity of a manganese oxide in the catalytic combustion of ketones. *Appl. Catal. B Environ.* **2001**, *33*, 1–8. [CrossRef]
11. Paulis, M.; Gandia, L.M.; Gil, A.; Sambeth, J.; Odriozola, J.A.; Montes, M. Influence of the surface adsorption–desorption processes on the ignition curves of volatile organic compounds (VOCs) complete oxidation over supported catalysts. *Appl. Catal. B Environ.* **2000**, *26*, 37–46. [CrossRef]
12. Álvarez-Galván, M.C.; de la Peña O'Shea, V.A.; Arzamendi, G.; Pawelec, B.; Gandía, L.M.; Fierro, J.L.G. Methyl ethyl ketone combustion over La-transition metal (Cr, Co, Ni, Mn) perovskites. *Appl. Catal. B Environ.* **2009**, *92*, 445–453. [CrossRef]
13. Tsou, J.; Magnoux, P.; Guisnet, M.; Orfao, J.J.M.; Figueiredo, J.L. Catalytic oxidation of volatile organic compounds: Oxidation of methyl-isobutyl-ketone over Pt/zeolite catalysts. *Appl. Catal. B Environ.* **2005**, *57*, 117–123. [CrossRef]
14. Li, W.B.; Wang, J.X.; Gong, H. Catalytic combustion of VOCs on non-noble metal catalysts. *Catal. Today* **2010**, *148*, 81–87. [CrossRef]
15. Papaefthimiou, P.; Ioannides, T.; Verykios, X. Combustion of non-halogenated volatile organic compounds over group VIII metal catalysts. *Appl. Catal. B Environ.* **1997**, *13*, 175–184. [CrossRef]
16. Liotta, L.F. Catalytic oxidation of volatile organic compounds on supported noble metals. *Appl. Catal. B Environ.* **2010**, *100*, 403–412. [CrossRef]
17. Ordóñez, S.; Bello, L.; Sastre, H.; Rosal, R.; Fernando, V.D. Kinetics of the deep oxidation of benzene, toluene, *n*-hexane and their binary mixtures over a platinum on γ-alumina catalyst. *Appl. Catal. B Environ.* **2002**, *38*, 139–149. [CrossRef]
18. Barresi, A.A.; Baldi, G. Deep catalytic oxidation of aromatic hydrocarbon mixtures: Reciprocal inhibition effects and kinetics. *Ind. Eng. Chem. Res.* **1994**, *33*, 2964–2974. [CrossRef]
19. Rusu, A.O.; Dumitriu, E. Destruction of volatile organic compounds by catalytic oxidation. *Environ. Eng. Manag. J.* **2003**, *2*, 273–302. [CrossRef]

20. Papaefthimiou, P.; Ioannides, T.; Verykios, X.E. Catalytic incineration of volatile organic compounds present in industrial waste streams. *Appl. Therm. Eng.* **1998**, *18*, 1005–1012. [CrossRef]
21. Centi, G. Supported palladium catalysts in environmental catalytic technologies for gaseous emissions. *J. Mol. Catal. A Chem.* **2001**, *173*, 287–312. [CrossRef]
22. Forzatti, P.; Lietti, L. Catalyst deactivation. *Catal. Today* **1999**, *52*, 165–181. [CrossRef]
23. Genty, E.; Brunet, J.; Pequeux, R.; Capelle, S.; Siffert, S.; Cousin, R. Effect of Ce substituted hydrotalcite-derived mixed oxides on total catalytic oxidation of air pollutant. *Mater. Today Proc.* **2016**, *3*, 277–281. [CrossRef]
24. Hibino, T.; Tsunashima, A. Formation of spinel from a hydrotalcite-like compound at low temperature: Reaction between edges of crystallites. *Clays Clay Miner.* **1997**, *45*, 842–853. [CrossRef]
25. Bera, P.; Rajamathi, M.; Hegde, M.S.; Kamath, P.V. Thermal behaviour of hydroxides, hydroxysalts and hydrotalcites. *Bull. Mater. Sci.* **2000**, *23*, 141–145. [CrossRef]
26. Rives, V. (Ed.) *Layered Double Hydroxides: Present and Future*; Nova Science Publishers Inc.: Hauppauge, NY, USA, 2001; ISBN 1590330609.
27. McCullagh, E.; Rigas, N.C.; Gleaves, J.T.; Hodnett, B.K. Selective oxidation of butan-2-one to diacetyl over vanadium pentoxide. An investigation by temporal analysis of products. *Appl. Catal. A Gen.* **1993**, *95*, 183–195. [CrossRef]
28. McCullagh, E.; McMonagle, J.B.; Hodnett, B.K. Kinetic study of the selective oxidation of butan-2-one to diacetyl over vanadium phosphorus oxide. *Appl. Catal. A Gen.* **1993**, *93*, 203–217. [CrossRef]
29. Santos, V.P.; Carabineiro, S.A.C.; Tavares, P.B.; Pereira, M.F.R.; Órfão, J.J.M.; Figueiredo, J.L. Oxidation of CO, ethanol and toluene over TiO$_2$ supported noble metal catalysts. *Appl. Catal. B Environ.* **2010**, *99*, 198–205. [CrossRef]
30. Lahousse, C.; Bernier, A.; Grange, P.; Delmon, B.; Papaefthimiou, P.; Ioannides, T.; Verykios, X. Evaluation of γ-MnO$_2$ as a VOC removal catalyst: Comparison with a noble metal catalyst. *J. Catal.* **1998**, *178*, 214–225. [CrossRef]
31. Beauchet, R.; Mijoin, J.; Batonneau-Gener, I.; Magnoux, P. Catalytic oxidation of VOCs on NaX zeolite: Mixture effect with isopropanol and o-xylene. *Appl. Catal. B Environ.* **2010**, *100*, 91–96. [CrossRef]
32. Beauchet, R.; Mijoin, J.; Magnoux, P. Improved catalytic oxidation of cumene by formation of catalytically active species during reaction over NaX zeolite. *Appl. Catal. B Environ.* **2009**, *88*, 106–112. [CrossRef]
33. Lars, S.; Andersson, T. Reaction networks in the catalytic vapor-phase oxidation of toluene and xylenes. *J. Catal.* **1986**, *98*, 138–149. [CrossRef]

© 2018 by the authors. Licensee MDPI, Basel, Switzerland. This article is an open access article distributed under the terms and conditions of the Creative Commons Attribution (CC BY) license (http://creativecommons.org/licenses/by/4.0/).

Article

Catalytic Activity Studies of Vanadia/Silica–Titania Catalysts in SVOC Partial Oxidation to Formaldehyde: Focus on the Catalyst Composition

Niina Koivikko [1,*], Tiina Laitinen [1], Anass Mouammine [1,2], Satu Ojala [1] and Riitta L. Keiski [1]

[1] Environmental and Chemical Engineering (ECE), Faculty of Technology, University of Oulu, P.O. Box 4300, FI-90014 Oulu, Finland; tiina.laitinen@oulu.fi (T.L.); mouammine@gmail.com (A.M.); satu.ojala@oulu.fi (S.O.); riitta.keiski@oulu.fi (R.L.K.)
[2] Laboratory of Catalysis and Corrosion of Materials (LCCM), Department of Chemistry, Faculty of Sciences, University of Chouaïb Doukkali, 20 Route de Ben Maachou, 24000 El Jadida, Morocco
* Correspondence: niina.koivikko@oulu.fi; Tel.: +358-(0)50-350-4188

Received: 31 December 2017; Accepted: 29 January 2018; Published: 2 February 2018

Abstract: In this work, silica–titania supported catalysts were prepared by a sol–gel method with various compositions. Vanadia was impregnated on SiO_2-TiO_2 with different loadings, and materials were investigated in the partial oxidation of methanol and methyl mercaptan to formaldehyde. The materials were characterized by using N_2 physisorption, X-ray diffraction (XRD), X-ray fluorescence spectroscopy (XRF), X-ray photoelectron spectroscopy (XPS), Scanning transmission electron microscope (STEM), NH_3-TPD, and Raman techniques. The activity results show the high importance of an optimized SiO_2-TiO_2 ratio to reach a high reactant conversion and formaldehyde yield. The characteristics of mixed oxides ensure a better dispersion of the active phase on the support and in this way increase the activity of the catalysts. The addition of vanadium pentoxide on the support lowered the optimal temperature of the reaction significantly. Increasing the vanadia loading from 1.5% to 2.5% did not result in higher formaldehyde concentration. Over the 1.5%V_2O_5/SiO_2 + 30%TiO_2 catalyst, the optimal selectivity was reached at 415 °C when the maximum formaldehyde concentration was ~1000 ppm.

Keywords: vanadium pentoxide; titanium dioxide; silicon dioxide; utilization of VOC; oxidative dehydrogenation; oxidative desulfurization

---

1. Introduction

Silicon dioxide SiO_2 and titanium dioxide TiO_2 support materials are used extensively in academic research and in industrial applications. Mixed SiO_2-TiO_2 materials have attracted many researchers, because it has been shown that mixed SiO_2-TiO_2 materials can provide certain advantages over single oxides. Benefits such as stronger metal–support interactions, higher acidity compared to single oxides, better resistance to sintering, and resistance against sulfur poisoning have been observed in earlier studies [1–7].

The sol–gel preparation procedure, not only being a rather easy way to prepare catalyst materials, provides advantages over the SiO_2-TiO_2 preparation process [6,8,9]. The sol–gel preparation steps include sol (colloid suspension) preparation, alkoxysilane hydrolysis, and condensation of silica and titania precursors. This leads to a structure where Ti is homogeneously dispersed in the silica matrix, which often results in better activity in oxidation applications. The presence of only a small amount of Ti leads to structural changes of the support as it causes narrowing of the Si-O bond length distribution. The support has a disordered tetrahedral structure in which Ti atoms are incorporated into the silica network [5]. Due to the advantages of silica–titania supports, they are used as supports for vanadia-containing catalysts in different applications [10–18].

This work focuses on the utilization of sulfur-containing volatile organic compound (SVOC) emissions. These emissions raise a lot of discussion, especially in the pulp and paper industry, as they are very odorous at low concentrations and their emission levels typically vary. Nowadays, methanol-containing streams are collected from the process and used as an energy source [19]. To increase the fuel value of methanol, the streams are concentrated and then directed to the combustion. Since the streams are utilized in the pulping process as an energy source, the formation of carbon dioxide emissions cannot be avoided. New approaches are needed to minimize the environmental load originating from methanol emissions and to utilize the stream in the production of new valuable chemicals more efficiently.

Compounds such as methyl mercaptan (MM), dimethyl sulphide (DMS), dimethyl disulphide (DMDS) and sulfur dioxide (SO_2) are present in mostly methanol (MeOH) and water containing stripper overhead gases (SOG) of the pulp mills [19,20]. In this approach, the MM containing methanol is considered as the raw material for formaldehyde production. We have previously shown the good activity of V_2O_5/SiO_2-TiO_2 in this application [21], and thus the aims of the current study lie in optimizing the catalyst composition and in finding more information from the catalyst properties affecting the formation of formaldehyde and possible other products.

Methanol oxidative dehydrogenation to formaldehyde over different vanadia catalysts has been under intensive investigation during recent decades [15,22–29]; however, the utilization of the contaminated methanol by sulfur compounds, such as methyl mercaptan, has been less studied. It has been stated that the performance of the materials depends on the surface structure of the vanadium on the chosen support. According to current knowledge, the activity of the vanadium catalysts is largely due to the presence of VO_4 sites on the supporting materials [30].

2. Results and Discussion

2.1. Characterization of Catalysts

Specific surface areas, pore volumes, and pore sizes of all fresh supports and catalysts are given in Table 1 with the catalyst elemental analysis results measured by X-ray fluorescence spectroscopy (XRF). In addition, the vanadia surface loadings are given for the catalysts.

Table 1. The Brunauer-Emmett–Teller Barret-Joyner-Halenda (BET-BJH) method and X-ray fluorescence spectroscopy (XRF) results calculated as oxides for the supports and vanadium-containing catalysts.

Support	Surface Area [$m^2 g^{-1}$]	Total Pore Volume [$cm^3 g^{-1}$]	Pore Size [nm]	XRF [%] V_2O_5	SiO_2	TiO_2	Surface Loading V_{atom} nm^{-2}
Si	225	0.1	1.9	-	~100	-	-
Ti	10	0.02	6.8	-	-	~100	-
SiTi(10)	560	0.28	2.0	-	70	9	-
SiTi(30)	590	0.35	2.3	-	57	27	-
SiTi(60)	220	0.11	2.1	-	35	56	-
Catalyst							
1.5V/Si	140	0.06	1.9	1.56	98	-	0.74
1.5V/Ti	10	0.01	7.3	1.44	-	98	9.54
0.75V/SiTi(30)	560	0.33	2.3	0.65	58	27	0.08
1.5V/SiTi(30)	500	0.3	2.3	1.7	66	31	0.23
2.5V/SiTi(30)	470	0.28	2.3	2.4	66	31	0.34

Pure silica (denoted as Si) has a much higher surface area and pore volume compared to pure TiO_2 (denoted as Ti), 225 $m^2 g^{-1}$ and 11 $m^2 g^{-1}$, respectively. Addition of titanium into silica leads to higher surface areas even with such low titania loadings as 10%. While titanium dioxide as a single oxide has a low surface area, the mixed metal oxide SiO_2-TiO_2 (SiTi) with the Si:Ti ratios of 90:10 and 70:30 shows notably higher surface area, for example over 550 $m^2 g^{-1}$. For SiTi(60) the surface area was ~220 $m^2 g^{-1}$ as a results of the higher Ti-content in the support. Generally, the surface areas of vanadia-containing catalysts were lower than surface areas of the corresponding supports due to partial blockage of the pores of the support by vanadia particles. This has also been observed by other researchers [7]. Pore sizes are on the level of ~2 nm for all silica containing catalysts. Titania supported catalysts have pore sizes between 6.8 nm and 7.3 nm. Based on the XRF results, the desired amount of V_2O_5 as well as the SiO_2-TiO_2 ratio were obtained by the sol–gel procedure rather well.

The X-ray diffraction (XRD) patterns of pure Ti, SiTi(10), SiTi(30) and 2.5%V/SiTi(30) (calcination at 500 °C) are shown in Figure 1. The increase in Ti concentration in the mixed SiTi catalysts results in an increase in the peak intensity in the XRD diffractogram. As silica is amorphous, the changes in the spectra are related to the amount of crystalline titania which are shown by more intense XRD peaks. Titania is mainly in the anatase form (an intense peak at 2θ = 25.28) with a crystallite size of approximately 20 nm in a pure TiO_2 sample. The crystallite size of anatase in SiTi(30) and 2.5%V/SiTi(30) was ~7 nm. The pure TiO_2 sample also showed the presence of small amounts of rutile-phase (7%), which was not visible in the mixed oxides. The XRD results did not show the peak corresponding to V_2O_5 (2θ = 20.26) with the V_2O_5-loading of 2.5%. This may be due to low vanadia loading, but it indicates also that vanadium oxide is probably present in a well-dispersed state on the SiTi(30) support. To be able to detect V_2O_5, for example, on the anatase form of titania, earlier studies have reported that the amount of vanadia should be over 5% or even higher [7,31].

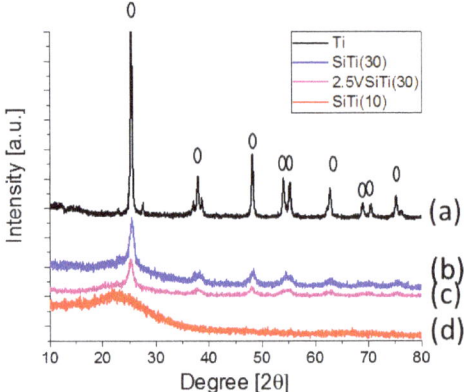

Figure 1. XRD patterns of the SiO_2-TiO_2 catalyst with different ratios of silica and titania (calcined at 500 °C): (**a**) Ti; (**b**) SiTi(30); (**c**) 2.5V/SiTi(30); (**d**) SiTi(10).

The X-ray photoelectron spectroscopy (XPS) analysis was done mainly to find out the oxidation degree of vanadium. Table 2 presents the binding energy (BE) values of V 2p, O 1s, Si 2p and Ti 2p for the fresh 1.5V/Si, 1.5V/Ti and 1.5V/SiTi(30). The charge correction was made by adjusting the main C 1s peak at 284.8 eV [32].

Table 2. Binding energies and full width at the half-maximum (FWHM) values presented in the brackets for the fresh 1.5%V_2O_5 catalysts supported on SiO_2, TiO_2 and SiTi(30).

Catalyst	V 2p [eV]		O 1s [eV]		Si 2p [eV]	Ti 2p [eV]	
	V 2p$_{3/2}$	V 2p$_{1/2}$				Ti 2p$_{1/2}$	Ti 2p$_{3/2}$
1.5V/Si	517.49 (1.86)	524.85 (3.37)	533.17 (1.78)	530.46 (1.58)	103.86 (1.66)	- -	- -
1.5V/Ti	517.62 (1.58)	524.94 (3.37)	531.92 (1.64)	530.41 (1.29)	- -	464.84 (2.13)	459.19 (1.27)
1.5V/SiTi(30)	517.64 (2.67)	525.06 (3.37)	532.88 (1.86)	530.43 (1.40)	103.60 (1.65)	465.00 (2.27)	459.20 (2.02)

The binding energy of the V 2p$_{3/2}$ core level depends on the oxidation state of the V cation; the curve fitting of V 2p$_{3/2}$ is often used to detect the different vanadium cation oxidation states present in vanadium oxide samples [32,33]. In this work, to define the oxidation states of vanadium, the data was collected together for V 2p and O 1s. The results are presented in Figure 2. For the V 2p, the spectra show a typical two-peak structure (V 2p$_{3/2}$ and V 2p$_{1/2}$) [33,34]. For each of the three catalysts the V 2p$_{3/2}$ spectra showed a peak at a binding energy value of ~517 eV (517.49–517.64 eV), indicating the oxidation state of V^{5+}. For the V 2p$_{1/2}$, spectra showed a peak at BE values between 524.8–525.0 eV, which also is connected to oxidation state V^{5+}. It was expected based on the literature that the vanadium could be present on the catalyst surface as V$_2$O$_5$ or a lower-valence vanadium oxide [32]. The typical value for full width at half-maximum (FWHM) for pure V$_2$O$_5$ is rather small (on the level of less than 2 eV). In our case this is consistent for Si and Ti supported vanadia. In the case of SiTi(30) support, the value is somewhat higher, indicating the possibility of the presence of other oxidation states of vanadium. The large widths of the V 2p lines might be also due to defects at the surface and/or disproportion at the surface [35].

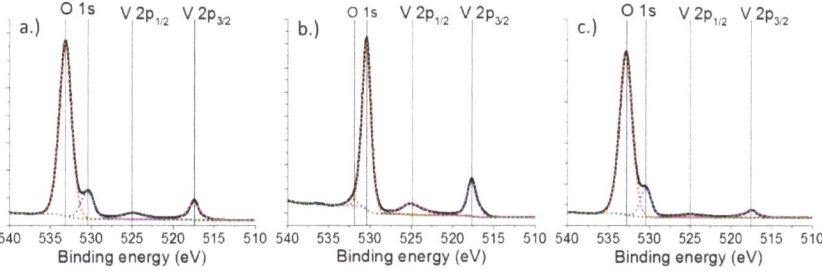

Figure 2. Electron spectra of (**a**) 1.5V/Si; (**b**) 1.5V/Ti; (**c**) 1.5V/SiTi(30) (calcined at 500 °C) studied with XPS.

The O 1s signal showed two clear peaks for silica containing samples: a major one at ca 532.9 eV (Si-O in the SiO$_2$ lattice) and a minor one at ca 530.3 eV (vanadium bonded oxygen) [12]. In addition, for the Si-containing samples the deconvolution of the Si 2p spectra revealed one peak at the binding energy values between 103.60–103.86 eV (see Table 2). Stakheev et al. [2] have observed earlier the presence of Si 2p peaks between 102.5–104.1 eV; that position depends on the added Ti amount in the sample.

The binding energy values for Ti 2p$_{3/2}$ remain approximately the same for both Ti-containing samples (~459.20 eV; see the Table 2). The addition of Ti to Si leads to a shift in the Si 2p$_{1/2}$ peak to a lower binding energy (total shift 0.26 eV). For the V/Ti catalyst the O 1s spectrum shows a peak centered at 530.41 eV and a tail extending to higher binding energy (centered at 531.92 eV). Odriozola et al. [36] explained that for the O 1s peak at ca. 532 eV, a so-called tail can be resolved and explained by the presence of hydroxyl species. Keränen et al. in 2003 [18] have also reported O 1s values of 530.0 and 531.6 eV for the V/Ti catalyst, which similarly explains the difference between the two peaks. For the SiTi support, in addition to the band maxima of O 1s observed at 532.88 eV (Si-O-Si) another band was observed at 530.43 eV, which is explained by either oxygen in Si-O-Ti structure or vanadium bonded oxygen [37]. Furthermore, the surface ratio of Si and Ti determined for 1.5%V/SiTi(30) was 80:20, which is very close to that determined by XRF for the bulk structure of the support. This gives an indication of the homogeneity of the prepared support.

The total acidities of SiTi materials with different ratios were determined with NH$_3$-TPD. The total amount of the desorbed ammonia representing the total amount of acid sites can be seen from Figure 3 (determined between 50 °C and 500 °C). Itoh et al. [1] have noted in 1974 that the acidity follows the trend of the surface area, but also that the total acidity is increased when SiO$_2$ is added to TiO$_2$. This is also expected based on the point of zero charge (PZC) values of the two oxides (pH at PZC for TiO_2 and SiO_2 are 6.0–6.4 and 2–4, respectively) [30]. In the same way, the current work shows that pure Ti has low acidity, which is increased after addition of silica ending up to pure SiO_2 that has the highest total acidity of the tested materials. The trend of total acidity based on NH_3-TPD is Si > SiTi(10) > SiTi(30) > SiTi(60) > Ti. Stakheev et al. in 1993 and Kobayashi et al. in 2005 [2,7] have presented results where the highest acidity was gained with the Si:Ti molar ratio of 1:1 determined by a titration method. This was not observed in our case. Considering the surface area of the samples, we did not notice clear correlation with the total acidity and the specific surface area of the samples, as shown in Figure 3. Concerning the strength of the acid sites, all the catalysts showed a broad low-temperature desorption peak (see Figure 4) centered at about 100 °C, which indicates the existence of mostly weak acid sites.

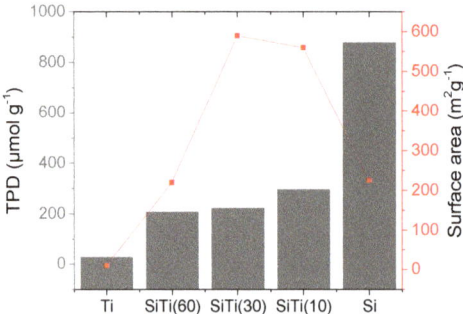

Figure 3. The results of the acidity of the silica and/or titania support materials measured with NH_3-TPD.

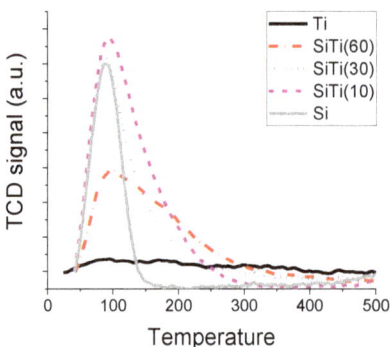

Figure 4. TPD profiles for ammonia desorption for silica and/or titania containing supports.

The STEM (Scanning transmission electron microscope) images of fresh catalysts, 1.5V/Si, 1.5V/Ti and 1.5V/SiTi(30), are presented in Figure 5. The STEM analysis was performed to investigate the microstructure of the catalysts. The EDS (Energy dispersive X-ray spectroscopy) mapping revealed the dispersions of vanadium on the catalysts' surface. On the silica support, vanadia forms bigger clusters (~250 nm) (see Figure 5a). This is due to the rather inert silica surface [38]. In the case of titanium dioxide and especially in the support of mixed SiTi, the particle size was much smaller and vanadium was better dispersed. Basic oxides usually exhibit good dispersion of vanadia and for that reason addition of less acidic titania to silica may result in better dispersion [39,40]. In the case of Ti support, the V_2O_5 is unevenly dispersed on the surface (see Figure 5b). The surface loading of V on TiO_2 is much higher (9.54 V_{atom} nm^{-2}) compared to the other catalysts, which have the surface loadings on the level of 0.08–0.74 V_{atom} nm^{-2}. This explains why vanadia is not well-dispersed on pure TiO_2 support.

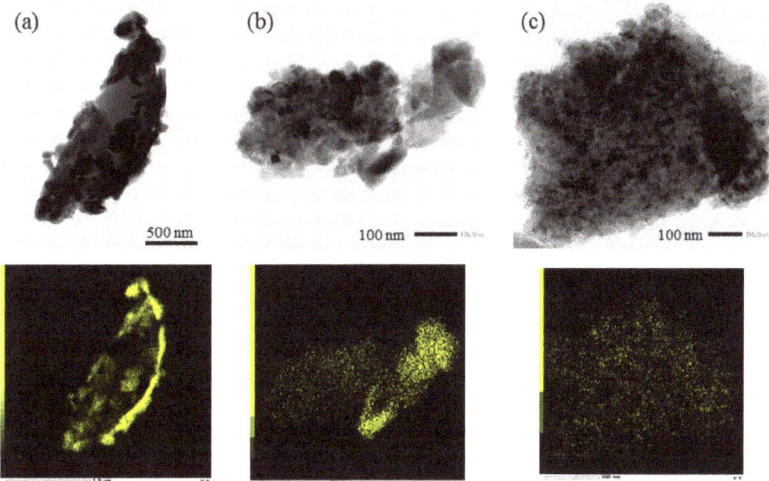

Figure 5. Scanning transmission electron microscope (STEM) images and Energy dispersive X-ray spectroscopy (EDS) mapping results of vanadia for fresh (**a**) 1.5V/Si (**b**) 1.5V/Ti (**c**) 1.5V/SiTi(30).

2.2. Results of Catalytic Oxidation Tests

The first series of the prepared catalysts and performance of the catalysts in the oxidation reaction of methanol (MeOH) and methyl mercaptan (MM) are presented in Figure 6a,b. In Figure 6a, the activity of the supports Si, Ti, and SiTi(30) are shown; and in Figure 6b, the effect of SiTi ratio on the gained formaldehyde concentration is presented. The earlier research results [21] show remarkably better activity of the support material in the oxidation of MeOH and MM as a mixture when Ti is added to Si. In this study, the main focus was on the investigation of the changes in activity to form formaldehyde with different Si:Ti ratios.

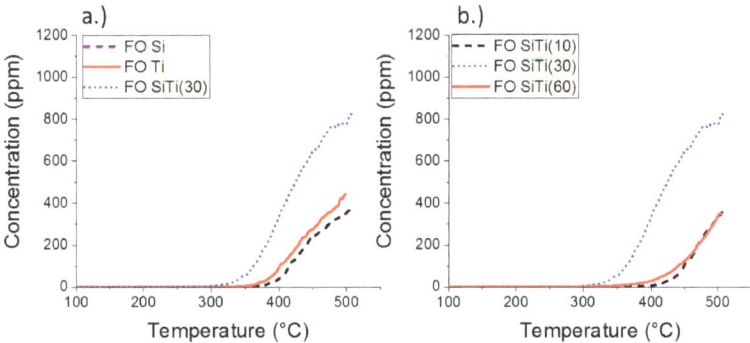

Figure 6. Formation of formaldehyde over (**a**) Si, Ti and SiTi(30) and (**b**) SiTi(10), SiTi(30) and SiTi(60) during the oxidation of methanol and methyl mercaptan (feed 500 ppm + 500 ppm, heating rate 5 °C min^{-1}).

As expected, with the addition of 30% titania to silica the formaldehyde production rose to more than double, compared to single oxide supports. What was unexpected (see Figure 6b) was that the SiTi ratio was crucial in gaining the highest formaldehyde concentration values. With the addition of 10% titania, no signs of activity improvement were observed. The 60% titania addition resulted only in ~400 ppm formaldehyde concentration, which is the same as with 10% Ti. Based on these results, it was decided to continue the activity testing with the SiTi(30) support as it was clearly the most active one from the tested materials with over 800 ppm formaldehyde production at 500 °C.

The addition of vanadia on the support was the second step in the current work. The 1.5% vanadium pentoxide addition increased the activity of all the tested catalysts. Results of the 1.5%V catalysts are presented in Figure 7. With single oxide silica and titania supports, the formaldehyde concentration rose to 1000 ppm, which was over 500 ppm more than with single oxide supports before vanadia addition. The addition of vanadia on SiTi(30) lowers the optimal reaction temperature significantly. For 1.5V/SiTi(30), the optimal temperature was already reached at around 400 °C.

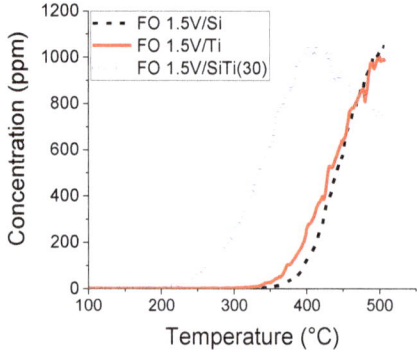

Figure 7. Formation of formaldehyde over pure silica or/and titania supports with 1.5% vanadium pentoxide during the oxidation of methanol and methyl mercaptan (feed 500 ppm + 500 ppm, heating rate 5 °C min^{-1}).

The results indicate that the good performance of single oxide Si or Ti supported catalysts requires an oxidation temperature over 500 °C—which we did not test—since lower temperatures are more interesting from an energy point of view, and because vanadium pentoxide as the catalyst material sets some limits to the activity testing. The melting point of V_2O_5 is 690 °C [41], and we wanted to stay at lower temperatures to minimize the sintering of vanadia during the reaction. Vanadium pentoxide as an active compound for MeOH and MM oxidation has been tested also at temperatures over 500 °C by Laitinen et al., 2016 [42].

As mentioned, both the Si:Ti ratio and the active compound V_2O_5 have a significant role in the reaction in reaching maximum formaldehyde production. The most active catalyst, 1.5V/SiTi(30), has the highest surface area and the best dispersion of vanadia based on the characterization. It has been shown that the dispersion of the active phase is dependent on the support composition. However, the results do not fully support the significance of the surface area, since SiTi(10) has higher specific surface area compared to single oxide silica, but it does not positively influence the activity results. On the other hand, the STEM results showed that the dispersion of the active phase is much better with the SiTi(30) support. Vanadia dispersion surely has a role in the activity of the material, but more testing with each of the materials with different vanadia particle sizes should be done. The effect of the total acidity in the activity is controversial. It is shown in the work of other researchers [1,7] that formaldehyde can be produced only with acidic oxides. However, the current results do not fully support these findings as pure silica resulted in the highest acidity of the tested materials, yet the lowest formaldehyde production. It seems that the acidity has to be at certain optimal level—not too high and not too low. In addition, the quality of acid sites (Brønsted-Lewis) may play a role, however we were not able to qualify the sites in the current study. Based on Tanabe et al. [43], SiO_2 contains mainly Brønsted acid sites (BAS) and TiO_2 Lewis acid sites (LAS). In the mixed oxide when TiO_2 is added to SiO_2, the amount of BAS in the material should increase. If this is true, the existence of both acid sites is important to be able to carry out the reactions under investigation. In addition, it has been also examined that the V/Ti catalysts show the presence of both BAS and LAS. The BAS are dominating acid sites in V_2O_2 [3]. The addition of vanadia to silica–titania support increases further the amount of BAS in the sample [44].

One of the objectives of this research was also to study the effect of the amount of vanadia impregnated on the catalysts. The results of SiTi(30) supported vanadia catalysts with different vanadia loadings (0.75%, 1.5% and 2.5%) are presented in Figure 8. As noted earlier, the addition of vanadia lowers the optimal temperature and the temperature where the reactions start. These current results reveal that the addition of vanadia from 1.5% to 2.5% does not result in increasing activity of the catalyst. Based on this finding we can expect that increasing the vanadia amount is not significant and it is reasonable to use a low amount of vanadia. This result is supported by Mouammine et al. [45], as in the case of V/Ti the increase in the V amount from 1.5% to 3% and to 10% did not result in higher formaldehyde yields.

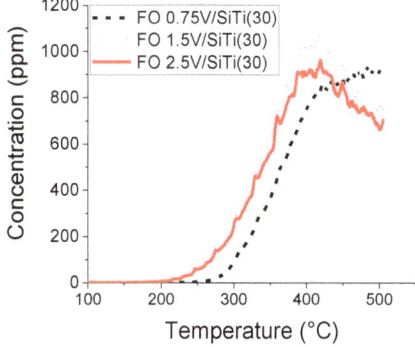

Figure 8. Formation of formaldehyde over silica–titania (V/SiTi(30)) catalysts with different V amounts (feed 500 ppm MeOH + 500 ppm MM, heating rate of 5 °C min^{-1}).

The main results from all the activity experiments are presented in Table 3. The table is presenting the key temperatures for each catalyst; the temperature in which the formation of the formaldehyde begins (A) and the temperature when the formaldehyde concentration reaches the maximum value during the experiment (B). From these values, it can be seen, that the formation of formaldehyde begins at the temperature level between 210 °C and 410 °C. The 1.5V/SiTi(30) catalyst has an optimal temperature of 415 °C (1060 ppm) and at 500 °C the formaldehyde concentration is only 760 ppm. This means that, after the optimal temperature, formaldehyde has reacted further to CO.

Table 3. Comparison of different catalysts in the reaction (key temperatures and gained maximum formaldehyde concentrations). Feed: 500 ppm methanol and 500 ppm methyl mercaptan, reaction temperature from room temperature to 500 °C.

Catalyst	Temperature A * [°C]	Temperature B # [°C]	Formaldehyde Concentration at Temp. B [ppm]	Formaldehyde Concentration at 500 °C [ppm]
SiO$_2$	380	500	360	360
TiO$_2$	360	500	440	440
SiTi(10)	410	500	330	330
SiTi(30)	315	500	780	780
SiTi(60)	365	500	340	340
0.75%V/SiTi(30)	215	480	930	918
1.5%V/SiTi(30)	215	415	1060	760
1.5%VSi	350	500	1030	1030
1.5%VTi	335	500	1000	1000
2.5%V/SiTi(30)	215	420	960	660

* The temperature when the formation of formaldehyde begins (over 10 ppm) during the activity test.
# The temperature when the formaldehyde concentration is highest during the test (optimal temperature for specific catalyst).

All the activity tests were repeated with the same catalyst twice. In Figure 9, the results of the repeated tests for 0.75V/SiTi(30), 1.5V/SiTi(30), and 2.5V/SiTi(30) are presented. The activity of the catalyst and the formation of formaldehyde follow exactly the same route in each test showing good repeatability of the test, but also giving some indications on the stability of the catalyst.

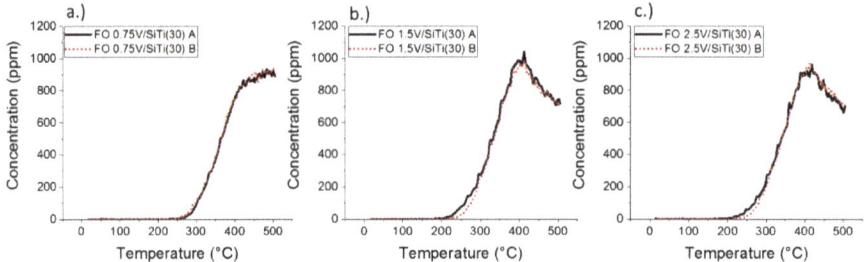

Figure 9. Formation of formaldehyde of the repeated tests for (**a**) 0.75V/SiTi(30); (**b**) 1.5V/SiTi(30) and (**c**) 2.5V/SiTi(30) (feed 500 ppm MeOH + 500 ppm MM, heating rate of 5 °C min^{-1}, A referring to test 1 and B referring to the repeated test 2).

In addition, more information on the stability of the 1.5%V catalyst with Si, Ti, and SiTi(30) supports was also revealed during the test at the optimal temperature for 8 h. The formaldehyde concentration remained unchanged through the whole experiment and no signs of activity loss were observed (figures not shown here).

Molecular structures of the dispersed vanadium oxide species on the support can be determined with Raman spectroscopy. The surface analysis was performed in order to investigate the structures of the fresh materials and if any changes after the 8 h stability test could be detected. The measurements were performed with Timegated® Raman device, in which the effect of the fluorescence is reduced from the resulting spectra. Figure 10 presents the Raman spectra of Si-, Ti-, SiTi(30)-supported V$_2$O$_5$ and bulk V$_2$O$_5$.

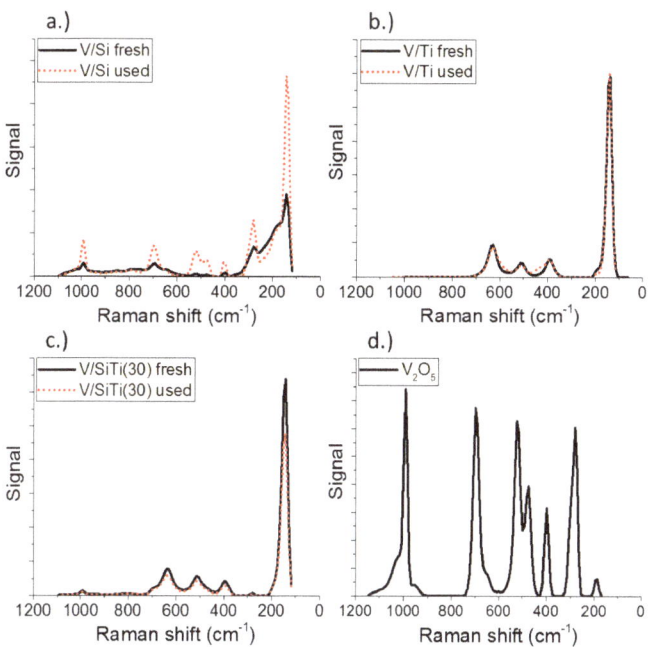

Figure 10. Timegated Raman spectra of (**a**) 1.5V/Si, (**b**) 1.5V/Ti, (**c**) 1.5V/SiTi(30) and (**d**) bulk V_2O_5.

Bulk vanadium pentoxide gives peaks at ~144, 188, 280, 398, 474, 524, 697, 940, 992 and 1030 cm^{-1}. The peak at 992 cm^{-1} is related to the symmetric stretching of V=O groups in the bulk vanadia [18,46] assigned as crystalline V_2O_5. It has been proposed that the less intense peaks at 940 cm^{-1} and 1030 cm^{-1} originate from isolated and polymerized surface vanadia species, respectively [10,47,48]. The intense peak at 700 cm^{-1} corresponds to lattice vibrations localized within the V-O-V bridge in the V_2O_5 structure.

For the Si-supported V_2O_5, similar features to crystalline V_2O_5 exist, since the observed peaks are in line with the peak positions of the bulk V_2O_5. For Ti-supported V_2O_5, the intense peaks from the TiO_2-support are visible at 142, 391, 510 and 633 cm^{-1} corresponding to the anatase TiO_2 [18]. In the case of both vanadia- and titania-containing samples the definite identification of the Raman spectra is difficult due to the proximity of intense crystalline V_2O_5 (~144 cm^{-1}) and TiO_2 anatase (~147 cm^{-1}) peaks. There were no peaks visible at 800–1000 cm^{-1}, which would be an indication that the dispersion of the isolated VO_x species on V/Ti was better than with V/Si. For the SiTi(30)-supported V_2O_5, the VO_x species at 278 and 991 cm^{-1} seem to be visible. Compared to the single oxide Ti support, a small peak at 991 cm^{-1} (V_2O_5) exists in the spectrum SiTi(30).

When the Raman spectra of the fresh catalysts and the catalysts after 8 h of testing were compared, no significant differences between the fresh and used samples were observed. The visible changes are only due to the changes in the signal intensity, which may in this case indicate more crystalline material (such as in the case of used VSi). This gives an indication on the stability of the material in presence of sulfur. More studies will be done in the future to find out details related to the oxidation of methanol in the presence of sulfur-containing compounds.

Figure 11 presents the formation of by-products over the 1.5V/SiTi(30) catalyst to study the performance of the catalyst in more detail. The main products were sulfur dioxide, carbon monoxide, and dimethyl disulphide, as expected based on the literature [19,49].

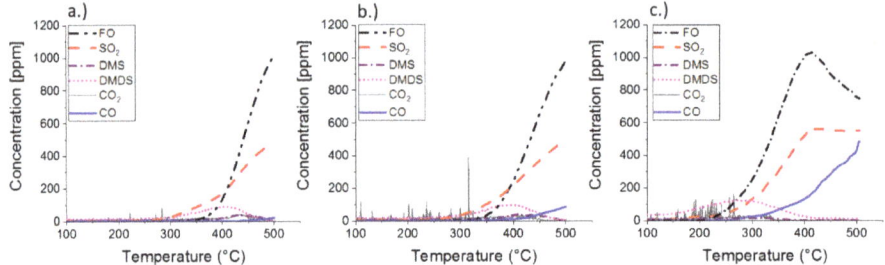

Figure 11. Formation of by-products in the oxidation of the mixture of methanol and methyl mercaptan (500 ppm + 500 ppm) over (**a**) 1.5V/Si; (**b**) 1.5V/Ti and (**c**) 1.5V/SiTi(30) catalysts (feed 500 ppm + 500 ppm, heating rate of 5 °C min^{-1}).

The formation of the products are results of the following reactions:

$$CH_3OH + 1/2O_2 \rightarrow CH_2O + H_2O \tag{1}$$

$$CH_3SH + 2O_2 \rightarrow CH_2O + SO_2 + H_2O \tag{2}$$

$$CH_3OH + 3/2O_2 \rightarrow CO_2 + 2H_2O \tag{3}$$

$$2CH_3SH + 1/2O_2 \rightarrow CH_3SSCH_3 + H_2O \tag{4}$$

$$CH_2O + 1/2O_2 \rightarrow CO + H_2O \tag{5}$$

$$CH_2O + O_2 \rightarrow CO_2 + H_2O \tag{6}$$

In the oxidation reaction of MM, sulfur dioxide is formed according to Reaction (2) and is the main product of the reaction together with formaldehyde. As formaldehyde reaches ~1000 ppm concentration in each of the tests, the formed SO$_2$ concentration is about 550 ppm in each presented case. The other reaction products, CO$_2$, H$_2$O and DMDS, are formed according to Reactions (3) and (4), which are complete oxidation of methanol and partial oxidation of MM to DMDS. The formation of DMDS is observed in the temperature range of 300 °C–450 °C for 1.5V/Si and 1.5V/Ti, reaching the maximum value approximately at 400 °C. For the SiTi supported catalyst the max. DMDS concentration of ~120 ppm is reached at temperature below 300 °C. The formation of carbon monoxide is detected when formaldehyde reacts further (Reactions (5) and (6)). This is clearly visible in the case of V/SiTi(30) at around 400 °C when the formaldehyde concentration starts to decrease and a significant amount of CO is formed. In other cases, the CO concentration stays at rather low levels (max. 85 ppm at 500 °C) as the concentration of formaldehyde is not decreasing during the temperature range used in the tests.

To compare the performance of the support and the vanadium containing catalyst, the concentrations of each gaseous product compound are presented in Figure 12. The comparison shown in the figure is done between the SiTi(30) support, marked as 1, and the 1.5V/SiTi(30) catalyst, marked as 2. To start with the reactants, there is a significant difference in the reaction when comparing the support and the catalyst. After vanadia addition on the support, the activity of the material in the reaction becomes significantly higher, which is visible in the changes in the reaction of MM. After vanadia addition, MM is already consumed starting at temperatures below 200 °C. Over the support, MM consumption begins at temperatures above 250 °C. MM conversion reached ~100% in all the tests. During the catalytic tests, the consumption of methanol begins at temperatures of around 300 °C. The differences in methanol partial oxidation activities between the support and the catalyst are not that significant, but the rate of the reaction is faster and higher methanol conversion is reached over the vanadia containing catalyst.

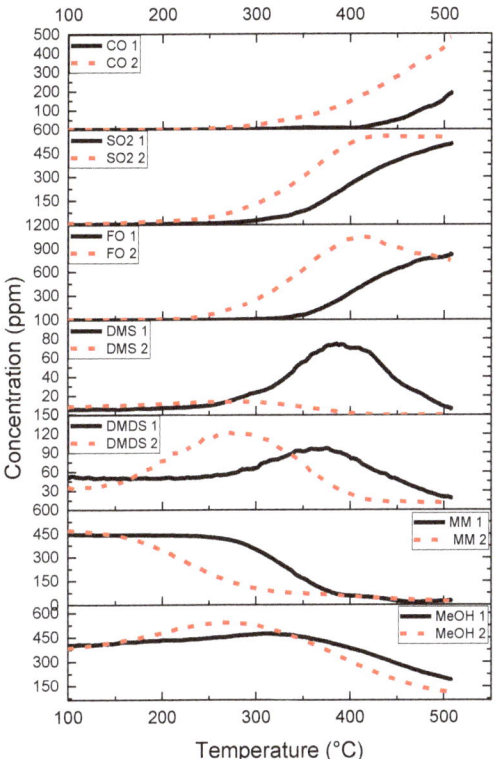

Figure 12. Formation of by-products in the oxidation of the mixture of methanol and methyl mercaptan (500 ppm + 500 ppm) over the SiTi(30) support(1) and 1.5V/SiTi(30) catalyst(2) (feed 500 ppm + 500 ppm, heating rate of 5 °C min^{-1}).

As explained earlier, the addition of vanadia on the SiTi support lowers the optimal temperature significantly, after which the formaldehyde concentration decreases when it starts to react further in the used reaction conditions. Figure 12 shows clearly the difference between the support and the catalyst in terms of the CO concentration, as in the presence of the VSiTi catalyst the CO concentration is more than twice the concentration reached over the SiTi support material. DMS concentration in both the experiments stays under 80 ppm, and for V/SiTi(30) the formation is insignificant. DMDS, for both the support and the vanadia containing catalyst, is observed as an intermediate product between 300–450 °C and 200–350 °C, respectively.

3. Materials and Methods

In this research, 5 different supports were prepared: pure SiO_2, TiO_2, and SiO_2-TiO_2 supports with the Si:Ti ratios of 90:10, 70:30, and 40:60 with no specific structure in target. After the optimization of the support composition, different amounts of vanadium pentoxide, 0.75%, 1.5%, and 2.5%, were impregnated on the support.

Mesoporous silica was prepared using the sol–gel method. A mass of 34.74 g of tetraethoxy orthosilicate (Si(OC$_2$H$_5$)$_4$, TEOS, 98% Sigma-Aldrich, St. Louis, MO, USA) was added to 54 g of absolute ethanol (Merck, Darmstadt, Germany) with a molar ratio of 1:7. A few drops of nitric acid were added to the solution to catalyze the hydrolysis step of the preparation. After dissolution of the silicon precursor, 24.5 g of ultrapure water (Sigma Aldrich, St. Louis, MO, USA), with a molar ratio of 1:8, was added dropwise. The obtained sol was then left for 15 days for aging. The TiO_2-support was synthesized using the sol–gel method. A Ti-containing solution was prepared first by adding 43 g titanium butoxide (Ti(OBu)$_4$, TBOT, 97% Sigma-Aldrich, St. Louis, MO, USA) to 83 g absolute ethanol with a molar ratio of 1:14 and stirred until complete dissolution of the Ti precursor. Then, ultra-pure water was added drop by drop until a molar ratio of 1:15 was reached.

The titania doped silica supports were prepared by dissolving an appropriate amount of both titanium and silicon precursors in absolute ethanol (molar ratio 1:14) to obtain a composition of $SiO_2(1-x)TiO_2(x)$ (with x = 0.1; 0.3; 0.6). A few drops of nitric acid were added to the solution, and then ultra-pure water was added with a molar ratio of 1:15. The solution was kept under stirring for 2 h. The final support was obtained after drying the gel at 90 °C overnight, and the dried gel was calcined at 500 °C for 2 h.

Addition of the active phase was done by using a wet impregnation method. At room temperature the appropriate amount of the vanadium precursor, vanadyl acetylacetonate (VO(acac)$_2$, 98% Sigma-Alrich, St. Louis, MO, USA) was dissolved in methanol. The support was added to the solution and kept under mechanical stirring overnight. The final catalysts were obtained after drying on a sand bath at 80 °C, and calcining at 500 °C for 2 h. The target amounts of vanadium pentoxide V_2O_5 on the support were 0.75 wt %, 1.5 wt %, and 2.5 wt %.

The materials prepared for this study and the abbreviations used in this article are presented in Table 4.

Table 4. Catalysts prepared for the oxidation studies.

Support/Catalyst	Abbreviation
SiO_2	Si
TiO_2	Ti
SiO_2 + 10%TiO_2	SiTi(10)
SiO_2 + 30%TiO_2	SiTi(30)
SiO_2 + 60%TiO_2	SiTi(60)
0.75%V_2O_5/ SiO_2+30%TiO_2	0.75V/SiTi(30)
1.5%V_2O_5/SiO_2	1.5V/Si
1.5%V_2O_5/TiO_2	1.5V/Ti
1.5%V_2O_5/SiO_2 + 30%TiO_2	1.5V/SiTi(30)
2.5%V_2O_5/SiO_2 + 30%TiO_2	2.5V/SiTi(30)

3.1. Characterization of Materials

The catalytic materials prepared for this study were characterized using different analytical techniques. The **BET-BJH** (Brunauer-Emmett–Teller Barret-Joyner-Halenda) method was used to determine the specific surface areas, pore volumes, and pore sizes of all the prepared materials. Nitrogen adsorption–desorption isotherms were recorded at −196 °C using an ASAP 2020 Micrometrics apparatus (Norcross, GA, USA). Before N_2 adsorption, the samples were degassed at 300 °C and kept under vacuum for 2 h.

X-ray fluorescence analyses **(XRF)** were carried out to study the V, Si, and Ti amounts on the catalyst samples. 0.2 g of the studied sample was mixed with 8.5 g of flux in a Pt-Au crucible followed by melting in an Eagon 2 furnace at 1150 °C. The analysis was performed with an Axios mAX X-ray fluorescence spectrometer (PANalytical, Almelo, The Netherlands).

X-ray diffractometer **(XRD)** Siemens D5000 was used to characterize the crystalline structure of the catalyst materials. Analysis data was recorded between 10° and 80°, with a step of 0.040°. The crystallite size of the active phase and support was estimated using the Scherrer formula:

$$D = \frac{k\lambda}{\beta_c \times \cos\theta}$$

where k is the shape factor ($k = 0.94$), λ is the wavelength of X-ray, θ is the Bragg angle, and β_c is the corrected line broadening defined as FWHM (full width at half maximum).

X-ray photoelectron spectroscopy (**XPS**) analysis was carried out by a Thermo Fischer Scientific ESCALAB 250Xi instrument (Waltham, MA, USA) using Al Kα (1486.6 eV) radiation to excite photoelectrons. The binding energy was normalized with respect to the position of the C1s peak at 284.8 eV. The XPS analysis was performed on 3 fresh catalyst samples: 1.5%V/Si, 1.5%V/Ti and 1.5%V/SiTi(30). The samples were put on an indium substrate and placed inside a vacuum chamber. Thermo Avantage software (v5.957, Thermo Fisher Scientific Inc., Waltham, MA, USA) was used in data analysis. Smart background subtraction was used and the spectra of O 1s, C 1s, Si 2p, Ti 2p and V 2p were recorded.

Temperature-programmed desorption measurements of ammonia (**NH$_3$-TPD**) were carried out by AutoChem II 2920 equipment (Micromeritics Instrument Corp., Norcross, GA, USA). A powder-form catalyst sample was placed inside the reactor and pretreated in a Helium flow (50 cm^3 min^{-1}) from room temperature to 500 °C at the rate of 5 °C min^{-1} for 30 min. The sample was then cooled to room temperature (He 50 cm^3 min^{-1}). The adsorption of 15% NH$_3$ in He (50 cm^3 min^{-1}) was carried out at RT for 60 min, and then the sample was flushed with He for 30 min to remove any physisorbed ammonia. TPD was performed in a He flow by raising the temperature to 550 °C with a rate of 5 °C min^{-1}. Sample amount of 100 mg was used for Si-containing catalysts and 180 mg for pure Ti due to significantly lower surface area. The desorbed amount of NH$_3$ was analyzed by a TCD detector. The area between 40 °C and 500 °C was used for determination of the total acidity of the samples.

Scanning transmission electron microscope (**STEM**) studies were carried out to analyze the particle size and the distribution of the vanadium pentoxide on the surface of the supports. The measurements were done using a JEOL JEM-2200FS apparatus (JEOL Ltd., Tokyo, Japan). The acceleration voltage of 200 kV was used. For the measurements, the catalyst samples were dispersed on a copper grid with ethanol. The apparatus was equipped with an Energy-Dispersive X-ray Spectrometer (EDS) apparatus (JEOL Dry SD100GV). The EDS was used to identify the absorbed chemical elements on the fresh catalysts. The measured catalysts were fresh 1.5V/Si, 1.5V/Ti and 1.5V/SiTi(30).

The **Raman** spectra were collected with a Timegate® Raman Spectrometer (Oulu, Finland) with a pulsed 532 nm fiber coupled laser and a rapid SPAD-detector. The data were collected with a Raman shift range from 150 to 1150 cm^{-1}. Data was curve fitted and analyzed with a Shsqui Matlab based software (v0.981, Timegate Instruments Oy, Oulu, Finland). The fresh 1.5V/Si, 1.5V/Ti and 1.5V/SiTi30 were measured and the measurement was repeated after the 8 h stability tests.

3.2. Catalytic Oxidation Tests

The catalytic partial oxidation tests were performed in a laboratory scale tubular quartz reactor. The experimental set-up presented in our previous work [21] was modified for further studies with a few basic improvements. The current set-up is presented in Figure 13. All the gas lines after the evaporator (heated to ~70 °C) were heated to 180 °C to avoid the condensation of the evaporated compound on the tube surfaces. Some condensation of methanol may still occur as the reactor part of the set-up is in the room temperature in the beginning of each test. Relatively high GHSV (~94,000 h^{-1}) was used in the experiments and the experimental procedure was following the same path as presented in the previous work [21].

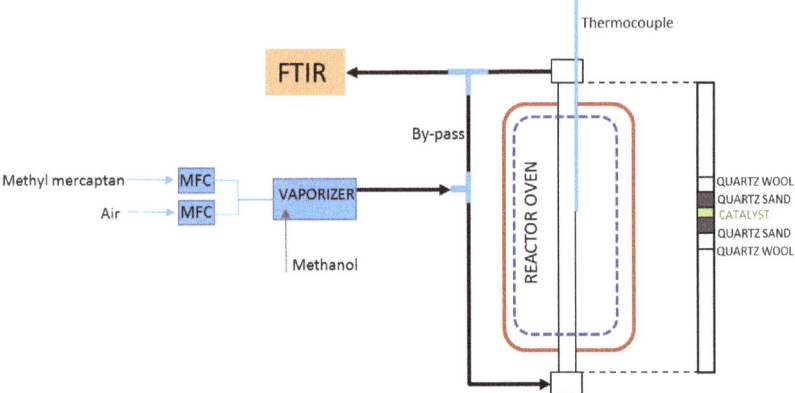

Figure 13. The experimental set-up for the catalyst activity testing.

All the experiments were conducted with 100 mg of a catalyst in the powder form. The feed concentration in each test was 500 ppm of methanol (Merck, Darmstadt, Germany) and 500 ppm of methyl mercaptan (Oy AGA Ab, Espoo, Finland) in a mixture. The reaction temperature was raised from room temperature to 500 °C with a heating rate of 5 °C min^{-1}. The outlet gas composition was analyzed by using a Gasmet FTIR Cr-2000 (Helsinki, Finland) analyzer equipped with an MCT detector. The following compounds were analyzed in each of the catalytic tests: carbon dioxide CO_2, carbon monoxide CO, nitrogen monoxide NO, nitrogen dioxide NO_2, nitrous oxide N_2O, sulfur dioxide SO_2, sulfur trioxide SO_3, methane CH_4, formaldehyde CH_2O, methyl mercaptan CH_4S, dimethyl sulfide C_2H_6S, dimethyl disulfide $C_2H_6S_2$, formic acid CH_2O_2, and methanol CH_4O. First the supports Si and Ti and mixed oxides SiTi(10), SiTi(30), and SiTi(60) were tested to find the possible changes in activity with respect to the Si:Ti ratio. The activity experiments were continued to study the effect of different vanadia loadings. The catalyst stability and repeatability of the catalytic tests were examined by testing the same catalyst in the same experimental conditions twice. The stabilities of 1.5 V catalysts were also tested in longer-term (8 h) experiments to gain some indication of the durability of the material. In 8 h tests, the catalyst amount of 200 mg was used. The same procedure as in the activity tests was used and the temperature was kept at the optimal formaldehyde production temperature (500 °C or lower) for 8 h.

4. Conclusions

V_2O_5 catalysts supported on SiO_2-TiO_2 have been characterized and tested in the oxidation of methanol and methyl mercaptan to formaldehyde. The results of the $V_2O_5/SiO_2 + TiO_2$ were compared with those obtained for V_2O_5/SiO_2 and V_2O_5/TiO_2 catalysts, showing that the composition of the support has a significant role in the catalytic behavior. The current oxidation results prove that silica–titania supported vanadia catalysts show good potential for use in the oxidation of sulfur contaminated methanol to formaldehyde. In the laboratory scale oxidation tests, significantly higher formaldehyde yields were achieved with the mixed silica–titania support. The important role of vanadium pentoxide in the catalyst is evident; the addition of the active compound results in higher activity of the catalysts, and lowers the optimal temperature of the reaction in which the desired partial oxidation products are formed. As expected, applying titania onto silica results in a higher surface area, which allows a good dispersion of vanadia. The acidity of the materials was dependent on the Si:Ti ratio of the support, but no solid conclusions about the role of the acidity in the activity of the catalysts in oxidation of methanol and methyl mercaptan could be made based on these findings. More detailed information on the surface of the catalysts and the reaction mechanisms is needed to optimize these materials further. For example, the role of the quality of the acid sites as well as the effect of the vanadia dispersion should be detailed.

Acknowledgments: This work was carried out with the financial support of the Academy of Finland (ELECTRA-project), the Emil Aaltonen Foundation, the Walter Ahlström Foundation, and the Tauno Tönning foundation. The authors also gratefully acknowledge the No-Waste project funding from the European Union Seventh Framework Programme (FP7), Marie Curie Actions under grant agreement no. PIRSES-GA-2012-317714. Jorma Penttinen, Kaisu Ainassaari, Markus Riihimäki, and Zouhair El Assal are acknowledged for their valuable help with the characterization of catalysts and Kirsi Ahtinen for her help with practical laboratory work. The Center of Microscopy and Nanotechnology of the University of Oulu is also acknowledged.

Author Contributions: N.K. designed the work presented in this article, performed most of the laboratory tests, and performed the literature search and wrote the paper. T.L. took part in experimental parts of this paper. She performed some parts of the material characterizations and took part in analysis of the activity measurements. A.M. took part in the material preparation phase of the work. S.O. was the main advisor of the work and took part in each phase of the work. R.L.K. is the principal supervisor of the thesis work of N.K. and thus has the main responsibility in all actions related to the manuscript.

Conflicts of Interest: The authors declare no conflict of interest.

References

1. Itoh, M.; Hattori, H.; Tanabe, K. The acidic properties of TiO_2-SiO_2 and its catalytic activities for the amination of phenol, the hydration of ethylene and the isomerization of butene. *J. Catal.* **1974**, *35*, 225–231. [CrossRef]
2. Stakheev, A.Y.; Shpiro, E.S.; Apijok, J. XPS and XAES study of TiO_2-SiO_2 mixed oxide system. *J. Phys. Chem.* **1993**, *97*, 5668–5672. [CrossRef]
3. Topsoe, N.Y.; Topsoe, H.; Dumesic, J.A. Vanadia/Titania Catalysts for Selective Catalytic Reduction (SCR) of Nitric-Oxide by Ammonia. *J. Catal.* **1995**, *151*, 226–240. [CrossRef]
4. Klein, S.; Thorimbert, S.; Maier, W.F. Amorphous Microporous Titania–Silica Mixed Oxides: Preparation, Characterization, and Catalytic Redox Properties. *J. Catal.* **1996**, *163*, 476–488. [CrossRef]
5. Walters, J.K.; Rigden, J.S.; Dirken, P.J.; Smith, M.E.; Howells, W.S.; Newport, R.J. An atomic-scale study of the role of titanium in TiO_2:SiO_2 sol–gel materials. *Chem. Phys. Lett.* **1997**, *264*, 539–544. [CrossRef]
6. Watson, R.B.; Ozkan, U.S. K/Mo Catalysts Supported over Sol–Gel Silica–Titania Mixed Oxides in the Oxidative Dehydrogenation of Propane. *J. Catal.* **2000**, *191*, 12–29. [CrossRef]
7. Kobayashi, M.; Kuma, R.; Masaki, S.; Sugishima, N. TiO_2-SiO_2 and V_2O_5/TiO_2-SiO_2 catalyst: Physico-chemical characteristics and catalytic behavior in selective catalytic reduction of NO by NH_3. *Appl. Catal. B Environ.* **2005**, *60*, 173–179. [CrossRef]
8. Morosanova, E.I. Silica and silica–titania sol–gel materials: Synthesis and analytical application. *Talanta* **2012**, *102*, 114–122. [CrossRef] [PubMed]
9. Nizar, U.K.; Efendi, J.; Yuliati, L.; Gustiono, D.; Nur, H. A new way to control the coordination of titanium (IV) in the sol–gel synthesis of broom fibers-like mesoporous alkyl silica–titania catalyst through addition of water. *Chem. Eng. J.* **2013**, *222*, 23–31. [CrossRef]
10. Jehng, J.M.; Wachs, I.E. The molecular structures and reactivity of V_2O_5/TiO_2/SiO_2 catalysts. *Catal. Lett.* **1992**, *13*, 9–19. [CrossRef]
11. Galán-Fereres, M.; Mariscal, R.; Alemany, L.J.; Fierro, J.L.G.; Anderson, J.A. Ternary V–Ti–Si catalysts and their behaviour in the CO + NO reaction. *J. Chem. Soc. Faraday Trans.* **1994**, *90*, 3711–3718. [CrossRef]
12. Quaranta, N.E.; Soria, J.; Cortés Corberán, V.; Fierro, J.L.G. Selective Oxidation of Ethanol to Acetaldehyde on V_2O_5/TiO_2/SiO_2 Catalysts. *J. Catal.* **1997**, *171*, 1–13. [CrossRef]
13. Reiche, M.A.; Ortelli, E.; Baiker, A. Vanadia grafted on TiO_2–SiO_2, TiO_2 and SiO_2 aerogels Structural properties and catalytic behaviour in selective reduction of NO by NH_3. *Appl. Catal. B Environ.* **1999**, *23*, 187–203. [CrossRef]
14. Gao, X.; Bare, S.R.; Fierro, J.L.G.; Wachs, I.E. Structural Characteristics and Reactivity/Reducibility Properties of Dispersed and Bilayered V_2O_5/TiO_2/SiO_2 Catalysts. *J. Phys. Chem. B* **1999**, *103*, 618–629. [CrossRef]
15. Burcham, L.J.; Deo, G.; Gao, X.; Wachs, I.E. In situ IR, Raman, and UV-Vis DRS spectroscopy of supported vanadium oxide catalysts during methanol oxidation. *Top. Catal.* **2000**, *11–12*, 85–100. [CrossRef]

16. Monaci, R.; Rombi, E.; Solinas, V.; Sorrentino, A.; Santacesaria, E.; Colon, G. Oxidative dehydrogenation of propane over $V_2O_5/TiO_2/SiO_2$ catalysts obtained by grafting titanium and vanadium alkoxides on silica. *Appl. Catal. A Gen.* **2001**, *214*, 203–212. [CrossRef]
17. Dias, C.R.; Portela, M.F.; Bañares, M.A.; Galán-Fereres, M.; López-Granados, M.; Peña, M.A.; Fierro, J.L.G. Selective oxidation of o-xylene over ternary V-Ti-Si catalysts. *Appl. Catal. A Gen.* **2002**, *224*, 141–151. [CrossRef]
18. Keränen, J.; Guimon, C.; Iiskola, E.; Auroux, A.; Niinistö, L. Atomic layer deposition and surface characterization of highly dispersed titania/silica-supported vanadia catalysts. *Catal. Today* **2003**, *78*, 149–157. [CrossRef]
19. Burgess, T.L.; Gibson, A.G.; Furstein, S.J.; Wachs, I.E. Converting waste gases from pulp mills into value-added chemicals. *Environ. Prog.* **2002**, *21*, 137–141. [CrossRef]
20. Wachs, I.E. Treating Methanol-Containing Waste Gas Streams. U.S. Patent 5907066, 6 March 2001.
21. Koivikko, N.; Laitinen, T.; Ojala, S.; Pitkäaho, S.; Kucherov, A.; Keiski, R.L. Formaldehyde production from methanol and methyl mercaptan over titania and vanadia based catalysts. *Appl. Catal. B Environ.* **2011**, *103*, 72–78. [CrossRef]
22. Wang, Q.; Madix, R.J. Partial oxidation of methanol to formaldehyde on a model supported monolayer vanadia catalyst: Vanadia on $TiO_2(1\ 1\ 0)$. *Surf. Sci.* **2002**, *496*, 51–63. [CrossRef]
23. Goodrow, A.; Bell, A.T. A theoretical investigation of the selective oxidation of methanol to formaldehyde on isolated vanadate species supported on titania. *J. Phys. Chem. C* **2008**, *112*, 13204–13214. [CrossRef]
24. Bronkema, J.L.; Bell, A.T. Mechanistic Studies of Methanol Oxidation to Formaldehyde on Isolated Vanadate Sites Supported on MCM-48 Mechanistic Studies of Methanol Oxidation to Formaldehyde on Isolated Vanadate Sites Supported on MCM-48. *J. Phys. Chem. C* **2007**, *111*, 420–430. [CrossRef]
25. Bronkema, J.L.; Bell, A.T. Mechanistic studies of methanol oxidation to formaldehyde on isolated vanadate sites supported on high surface area zirconia. *J. Phys. Chem. C* **2008**, *112*, 6404–6412. [CrossRef]
26. Vining, W.C.; Strunk, J.; Bell, A.T. Investigation of the structure and activity of $VO_x/ZrO_2/SiO_2$ catalysts for methanol oxidation to formaldehyde. *J. Catal.* **2011**, *281*, 222–230. [CrossRef]
27. Burcham, L.J.; Wachs, I.E. The origin of the support effect in supported metal oxide catalysts: In situ infrared and kinetic studies during methanol oxidation. *Catal. Today* **1999**, *49*, 467–484. [CrossRef]
28. Kropp, T.; Paier, J.; Sauer, J. Oxidative dehydrogenation of methanol at ceria-supported vanadia oligomers. *J. Catal.* **2017**, *352*, 382–387. [CrossRef]
29. Kim, M.H.; Ebner, J.R.; Friedman, R.M.; Vannice, M.A. Determination of Metal Dispersion and Surface Composition in Supported Cu–Pt Catalysts. *J. Catal.* **2002**, *208*, 381–392. [CrossRef]
30. Wachs, I.E. Catalysis science of supported vanadium oxide catalysts. *Dalt. Trans.* **2013**, *42*, 11762–11769. [CrossRef] [PubMed]
31. Chary, K.V.R.; Kishan, G.; Lakshmi, K.S.; Ramesh, K. Studies on dispersion and reactivity of vanadium oxide catalysts supported on titania. *Langmuir* **2000**, *16*, 7192–7199. [CrossRef]
32. Mendialdua, J.; Casanova, R.; Barbaux, Y. XPS studies of V_2O_5, V_6O_{13}, VO_2 and V_2O_3. *J. Electron Spectrosc. Relat. Phenom.* **1995**, *71*, 249–261. [CrossRef]
33. Silversmit, G.; Depla, D.; Poelman, H.; Marin, G.B.; De Gryse, R. Determination of the V2p XPS binding energies for different vanadium oxidation states (V^{5+} to V^{0+}). *J. Electron Spectrosc. Relat. Phenom.* **2004**, *135*, 167–175. [CrossRef]
34. Biesinger, M.C.; Payne, B.P.; Grosvenor, A.P.; Lau, L.W.M.; Gerson, A.R.; Smart, R.S.C. Resolving surface chemical states in XPS analysis of first row transition metals, oxides and hydroxides: Cr, Mn, Fe, Co and Ni. *Appl. Surf. Sci.* **2011**, *257*, 2717–2730. [CrossRef]
35. Demeter, M.; Neumann, M.; Reichelt, W. Mixed-valence vanadium oxides studied by XPS. *Surf. Sci.* **2000**, *454*, 41–44. [CrossRef]
36. Odriozola, J.A.; Soria, J.; Somorjai, G.A.; Heinemann, H.; de la Garcia Banda, J.F.; Lopez Granados, M.; Conesa, J.C. Adsorption of nitric oxide and ammonia on vanadia-titania catalysts: ESR and XPS studies of adsorption. *J. Phys. Chem.* **1991**, *95*, 240–246. [CrossRef]
37. Gao, X.; Bare, S.R.; Fierro, J.L.G.; Banares, M.A.; Wachs, I.E. Preparation and in-situ Spectroscopic Characterization of Molecularly Dispersed Titanium Oxide on Silica. *J. Phys. Chem. B* **1998**, *102*, 5653–5666. [CrossRef]

38. Gao, X.; Bare, S.R.; Weckhuysen, B.M.; Wachs, I.E. In Situ Spectroscopic Investigation of Molecular Structures of Highly Dispersed Vanadium Oxide on Silica under Various Conditions. *J. Phys. Chem. B* **1998**, *102*, 10842–10852. [CrossRef]
39. Deo, G.; Wachs, I.E. Predicting molecular structures of surface metal oxide species on oxide supports under ambient conditions. *J. Phys. Chem.* **1991**, *95*, 5889–5895. [CrossRef]
40. Blasco, T.; Nieto, J.M.L. Oxidative dyhydrogenation of short chain alkanes on supported vanadium oxide catalysts. *Appl. Catal. A Gen.* **1997**, *157*, 117–142. [CrossRef]
41. Rumble, J.R. (Ed.) *CRC Handbook of Chemistry and Physics*, 98th ed.; Internet Version 2018; CRC Press: Boca Raton, FL, USA, 2017.
42. Laitinen, T.; Ojala, S.; Koivikko, N.; Mouammine, A.; Keiski, R.L. Stability of a vanadium-based catalyst in the partial oxidation of a mixture of methanol and methyl mercaptan. In Proceedings of the 17th Nordic Symposium on Catalysis, Lund, Sweden, 14–16 June 2016.
43. Tanabe, K.; Sumiyoshi, T.; Shibata, K.; Kiyoura, T.; Kitagawa, J. A new hypothesis regarding the surface acidity of binary metal oxides. *Bull. Chem. Soc. Jpn.* **1974**, *47*, 1064–1066. [CrossRef]
44. Keränen, J.; Carniti, P.; Gervasini, A.; Iiskola, E.; Auroux, A.; Niinistö, L. Preparation by atomic layer deposition and characterization of active sites in nanodispersed vanadia/titania/silica catalysts. *Catal. Today* **2004**, *91–92*, 67–71. [CrossRef]
45. Mouammine, A.; Ojala, S.; Pirault-Roy, L.; Bensitel, M.; Keiski, R.; Brahmi, R. Catalytic partial oxidation of methanol and methyl mercaptan: Studies on the selectivity of TiO$_2$ and CeO$_2$ supported V$_2$O$_5$ catalysts. *Top. Catal.* **2013**, *56*, 650–657. [CrossRef]
46. Went, G.T.; Oyama, S.T.; Bell, A.T. Laser Raman spectroscopy of supported vanadium oxide catalysts. *J. Phys. Chem.* **1990**, *94*, 4240–4246. [CrossRef]
47. Vuurman, M.A.; Wachs, I.E.; Hirt, A.M. Structural determination of supported vanadium pentoxide-tungsten trioxide-titania catalysts by in situ Raman spectroscopy and X-ray photoelectron spectroscopy. *J. Phys. Chem.* **1991**, *95*, 9928–9937. [CrossRef]
48. Wachs, I.E.; Deo, G.; Weckhuysen, B.M.; Andreini, A.; Vuurman, M.A.; de Boer, M.; Amiridis, M.D. Selective Catalytic Reduction of NO with NH$_3$ over Supported Vanadia Catalysts. *J. Catal.* **1996**, *161*, 211–221. [CrossRef]
49. Reuss, G.; Disteldorf, W.; Gamer, A.O.; Hilt, A. Formaldehyde. In *Ulmann's Encyclopedia of Industrial Chemistry*; Wiley-VCH: Weinheim, Germany, 2012; Volume 15, pp. 735–768.

© 2018 by the authors. Licensee MDPI, Basel, Switzerland. This article is an open access article distributed under the terms and conditions of the Creative Commons Attribution (CC BY) license (http://creativecommons.org/licenses/by/4.0/).

Review

Ag/CeO$_2$ Composites for Catalytic Abatement of CO, Soot and VOCs

M. V. Grabchenko [1], N. N. Mikheeva [1], G. V. Mamontov [1], M. A. Salaev [1], L. F. Liotta [2],*
and O. V. Vodyankina [1],*

1. Laboratory of catalytic research, Tomsk State University, 36, Lenin Ave., Tomsk 634050, Russia; marygra@mail.ru (M.V.G.); natlitv93@yandex.ru (N.N.M.); grigoriymamontov@mail.ru (G.V.M.); mihan555@yandex.ru (M.A.S.)
2. Institute for the Study of Nanostructured Materials, Via Ugo La Malfa 153, 90146 Palermo, Italy
* Correspondence: leonarda.liotta@ismn.cnr.it (L.F.L.); vodyankina_o@mail.ru (O.V.V.); Tel.: +7-349-605-8182 (L.F.L.); +7-905-990-44-53 (O.V.V.)

Received: 23 June 2018; Accepted: 11 July 2018; Published: 16 July 2018

Abstract: Nowadays catalytic technologies are widely used to purify indoor and outdoor air from harmful compounds. Recently, Ag–CeO$_2$ composites have found various applications in catalysis due to distinctive physical-chemical properties and relatively low costs as compared to those based on other noble metals. Currently, metal–support interaction is considered the key factor that determines high catalytic performance of silver–ceria composites. Despite thorough investigations, several questions remain debating. Among such issues, there are (1) morphology and size effects of both Ag and CeO$_2$ particles, including their defective structure, (2) chemical and charge state of silver, (3) charge transfer between silver and ceria, (4) role of oxygen vacancies, (5) reducibility of support and the catalyst on the basis thereof. In this review, we consider recent advances and trends on the role of silver–ceria interactions in catalytic performance of Ag/CeO$_2$ composites in low-temperature CO oxidation, soot oxidation, and volatile organic compounds (VOCs) abatement. Promising photo- and electrocatalytic applications of Ag/CeO$_2$ composites are also discussed.

Keywords: silver–ceria; metal–support interaction; CO oxidation; soot oxidation; VOCs abatement

1. Introduction

Air pollution is a major environmental problem. According to the World health organization, ambient air pollution contributes to 6.7 percent of all deaths worldwide [1], and the emissions of harmful compounds from industrial plants and motor vehicles in crowded urban areas are getting more attention. By reducing the level of air pollution, countries can reduce the morbidity rates of heart disease, lung cancer, chronic and acute respiratory diseases, etc. Many substances cause air pollution, including carbon monoxide (CO), particulate matter, ozone, nitrogen dioxide, soot, sulfur dioxide, organic dyes, etc., with CO being the most common among these pollutants. Volatile organic compounds (VOCs) comprising organic compounds with an initial boiling point inferior or equal to 250 °C (measured at a standard pressure of 101.3 kPa) also impact pollution of indoor and outdoor air [2]. In a recent review [3], the authors consider several main classes of VOCs, including halogenated VOCs, aldehydes, aromatic compounds, alcohols, ketones, polycyclic aromatic hydrocarbons, etc.

Therefore, air cleaning is a pivotal challenge, and new solutions are required. Catalytic total oxidation of organic pollutants into CO$_2$ and water is the most effective way to address this challenge. Metal/ceria-based catalysts were found promising heterogeneous catalysts for CO, soot and VOCs oxidation, and the highly dispersed noble metals (Me = Au, Pt, Pd, Ru, etc.) were used as the active components of these catalysts. The ceria-supported catalysts containing Pd [4–11], Pt [12–16], Au [17–23], Ru [24,25], Rh [26–29], and Cu [30,31] were proposed. Metal oxide-based catalysts

(Co_3O_4 [32], MnO_x [33], etc.) also attracted wide interest. However, a significant part of the developed catalysts has limited use under real conditions due to high costs of noble metals (the loading of palladium, platinum, gold is of 2–10 wt. %), relatively low stability to "hard" conditions of oxidation processes causing the loss of active component and reduction of the catalyst activity and selectivity. Therefore, the development of high-performance affordable and stable catalysts for low-temperature total oxidation of harmful compounds is still challenging. Efficiency and costs of such catalysts are connected with proper selection of the type of active component, support, and preparation method [34].

Recently, supported silver catalysts have brought about wide interest due to their high activity in low-temperature oxidation processes. Different supports are studied (SiO_2, CeO_2, MnO_x, TiO_2, Al_2O_3, ZrO_2, etc.) [35–38]. It is shown that an enhanced catalytic activity of Ag-based catalysts can be achieved by using reducible metal oxides as supports and by controlling the metal–support interaction to provide synergistic effect between active sites of the support and noble metal [39].

Among the supports mentioned, CeO_2 brings about high interest, since it combines exceptional redox and acid-base properties with oxygen storage, which can be controlled by proper preparation methods and treatments. Moreover, these distinctive properties of ceria cause its wide applications as a support for catalysts. For this reason, highly active and relatively inexpensive Ag/CeO_2 composite is considered promising heterogeneous catalyst for total oxidation of harmful organic compounds, including formaldehyde [40], CO [41–43], soot [44–49]. The Ag/CeO_2 composites can also be used in photo- [50] and electrocatalysis [51,52], reduction of NO_x [41], methane oxidation reaction [53], preferential oxidation of CO in excess of H_2 (PROX CO) [54,55] as well as in biochemistry due to the bactericidal properties of both ceria and silver [56]. It is noteworthy also that these composites are applied in selective oxidation of organic compounds.

In this review, we provide a survey of the current state of catalytic total oxidation of CO, soot, and VOCs over Ag/CeO_2 catalysts. The reactions under consideration are discussed from the perspective of (1) morphology and size effects of both Ag and CeO_2 particles, including their defective structure, (2) chemical and charge state of silver, (3) charge transfer between silver and ceria, (4) role of oxygen vacancies, (5) reducibility of support and the catalyst on the basis thereof.

2. Topical Processes

2.1. CO Oxidation

CO oxidation is one of the most studied reactions in catalysis science. It is of great fundamental and practical importance, since CO is formed as a by-product in many industrially important oxidation reactions (e.g., methanol oxidation to formaldehyde [57], ethylene glycol oxidation to glyoxal [58], etc.). CO seriously affects the environment and human health [59].

Ceria-based catalysts are among the most promising materials for CO oxidation [59–61]. A comparison of ceria-supported noble metals shown that Pd/CeO_2 and Au/CeO_2 catalysts were more active in CO oxidation than Ag/CeO_2 [62,63]. However, the relatively low activity of Ag-containing catalysts in this case may be connected with non-optimal conditions of preparation and pre-treatment of Ag/CeO_2 catalyst. Ag/SiO_2 catalysts are known to be able to catalyze low-temperature CO oxidation even at temperatures below 0 °C [64–67]. Such factors as the size of Ag nanoparticles, the pre-treatment conditions of both support and catalyst, metal–support interaction determine the catalytic activity of silver catalysts in CO oxidation. In Ref. [68] it was shown that addition of CeO_2 to Ag/SiO_2 improved the catalytic activity in CO oxidation due to the cooperation of oxidative species on Ag and ceria. Thus, the study of Ag/CeO_2 catalysts deserves special attention to reveal the reasons for high catalytic activity and find the approaches to its regulation.

According to literature, the method of Ag/CeO_2 synthesis determines the catalytic properties. Thus, in Ref. [41] the 10% Ag/CeO_2 catalysts were prepared by impregnation and deposition–precipitation techniques. The catalysts prepared by impregnation demonstrated higher activity in CO and propylene oxidation. This finding was associated with formation of Ag^{2+} species in these catalysts, confirmed by Electron Paramagnetic Resonance (EPR). Such species improve the redox properties due to creation of three different redox couples: Ag^{2+}/Ag^+, Ag^{2+}/Ag^0, and Ag^+/Ag^0.

The effect of shape of ceria nanoparticles on the catalytic properties of ceria-based catalysts is also discussed in the review [69]. In Ref. [70] synthesis of ceria nanopolyhedra, nanorods, and nanocubes by a hydrothermal method is described (Figure 1). The oxygen storage capacity of CeO_2 nanorods and nanocubes was attributed to both surface and bulk oxygen species. The lowest oxygen storage capacity for ceria nanopolyhedra was attributed to a predominance of (111) boundaries on the surface of particles with low reaction ability toward CO. Thus, the shape-selective synthetic strategy may be used for designing the catalysts with desired oxidative activity.

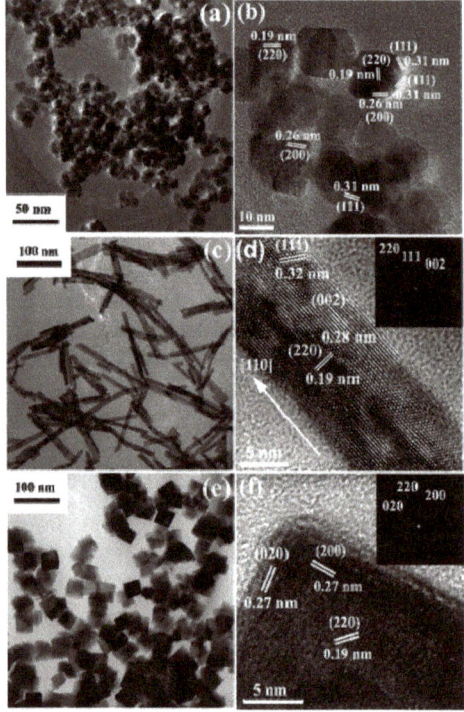

Figure 1. TEM (**a**) and HRTEM (**b**) images of CeO_2 nanopolyhedra. TEM (**c**) and HRTEM (**d**) images of CeO_2 nanorods, inset is a fast Fourier transform (FFT) analysis. TEM (**e**) and HRTEM (**f**) images of CeO_2 nanocubes, inset is a FFT analysis. Reproduced from Ref. [70] with the permission from ACS Publications.

In Ref. [71] the catalytic activity of ceria rods, cubes and octahedra was studied in CO oxidation. The highest activity of ceria nanorods was attributed to a predominance of (110) and (100) surfaces, while the lowest activity of ceria octahedra was caused by a predominance of (111) surface. The activity of different surfaces also depends on the energy of oxygen vacancy formation, which is predicted to follow the reverse order of lattice oxygen reactivity: (110) < (100) < (111). Supporting of silver on the surface of ceria nanoparticles with different shapes by conventional incipient wetness impregnation followed by calcination at 500 °C led to creation of additional oxygen vacancies in ceria surface [43]. Ag nanoparticles were suggested to facilitate the formation of oxygen vacancies in ceria surface in a larger extent than in case of positively charged Ag_n^+ clusters. Thus, Ag loading (1 and 3 wt. %) in Ag/CeO_2 affects the amount of Ag^0 and Ag_n^+ clusters that yields different concentrations of surface oxygen vacancies and, hence, different activity in CO oxidation. Ag^0 nanoparticles (NPs) promote the reducibility of surface lattice oxygen and catalytic activity of CeO_2 in CO oxidation. The control of the shape of CeO_2 may be used as a strategy to design the metal/CeO_2 catalysts with reduced amounts of noble metals. An increase of the Ag content from 1 to up to 3 wt. % mitigates the difference in turnover frequency (TOF) CO for the composites based on nanocubes and nanorods that allows concluding on the need of coexistence of charged Ag_n^+ species and reduced Ag^0 NPs on the CeO_2 surface to create an active catalyst.

The role of oxygen vacancies of Ag/CeO_2 catalysts in CO oxidation is also discussed in Ref. [72]. Using Raman spectroscopy, it was shown that Ag promoted the formation of oxygen vacancies in ceria. This effect is pronounced, when CeO_2 and Ag/CeO_2 were reduced in CO/N_2 atmosphere up to 300 °C (Figure 2a,b). Treatment in oxygen atmosphere leads to the decreased amount of oxygen vacancies (Figure 2c,d). Thus, the introduction of Ag into CeO_2 promotes the activation of lattice oxygen of ceria and formation of oxygen vacancies that is the main reason for enhanced catalytic activity of Ag/CeO_2 in CO oxidation.

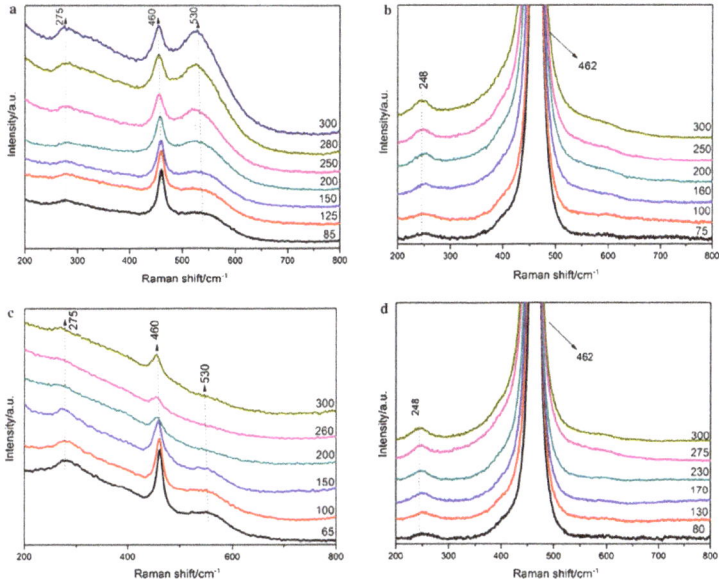

Figure 2. Raman spectra of different catalysts under different reaction conditions (a) Ag/CeO_2—5 vol% CO/N_2, (b) CeO_2—5 vol% CO/N_2, (c) Ag/CeO_2–O_2, (d) CeO_2–O_2. Reproduced from Ref. [72] with the permission from Springer.

The role of the shape of ceria nanoparticles in CO oxidation over Ag/CeO_2 was also discussed in terms of the complex or hierarchical structure of ceria. The Ag-based catalysts supported on mesoporous CeO_2 prepared by hard-template method and surfactant-template method was studied in CO oxidation in Ref. [42]. Mesoporous ceria was prepared by hard-template method using the SBA-15 material as a template, which was etched by NaOH. Hexadecyl trimethyl ammonium bromide (CTAB) was used as a classical soft template to synthesize ceria by surfactant-template method. Mesoporous ceria prepared by hard-template method was the preferable support for Ag catalysts, and total conversion of CO (200 mg catalyst, 1% CO, a gas flow of 30 mL/min) for this catalyst was achieved at 65 °C. High activity of this catalyst was attributed to oxygen vacancies in mesoporous CeO_2 support, which stabilizes dispersed silver and facilitates the transfer of electrons from Ag to CeO_2 via the Ag–CeO_2 interface. However, one cannot exclude the participation of SiO_2 used as a template to produce mesoporous CeO_2 in formation of Ag-containing species highly reactive toward low-temperature CO oxidation.

In Ref. [73] Ag/CeO_2 catalysts with the Ag loading from 5 to 20 wt. % were prepared by the HCl etching of $CuO/CeO_2/Ag_2O$ mixed oxides followed by CuO removal. The formation of Ag nanoparticles inside the ultrafine nanoporous CeO_2 with sizes of pore channels below 20 nm was observed after reduction by glucose in solution. The obtained composites also showed enhanced catalytic activity in CO oxidation in comparison with CeO_2–Ag composite prepared by co-precipitation method, and the highest catalytic activity was observed for catalysts with 10 wt. % loading of Ag ($T_{50\%} \approx 130\ °C$, 1% CO and 10% O_2, WHSV of 60,000 mL g^{-1} h^{-1}).

The CeO_2 mesoporous spheres with a diameter of ~100 nm and Ag catalysts on the basis thereof were synthesized in Ref. [74] (Figure 3). CeO_2 mesoporous spheres were synthesized using glycol as a solvent with addition of C_2H_5COOH in an autoclave at 180 °C for 200 min. Ag NPs were prepared separately, and their dispersion in cyclohexane was stirred together with CeO_2 mesoporous spheres. The catalysts were characterized by high surface area (216 m^2/g) and regular morphology. Ag molar content was 10%. CO conversion achieved 96.5% at 70 °C (100 mL/min) and the enhanced catalytic performance in CO oxidation was attributed to the unique structure of ceria support.

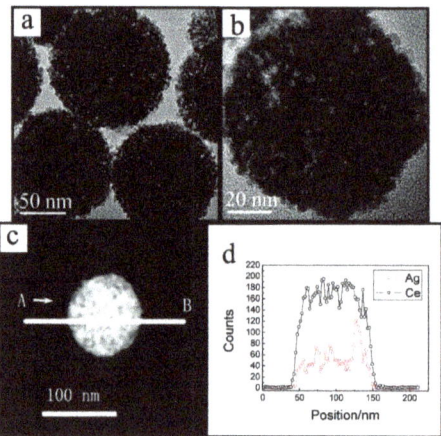

Figure 3. (**a** and **b**) TEM images with different magnifications of CeO_2 mesoporous spheres supported by a Ag nanoparticle catalyst. (**c**) Darkfield scanning TEM image of a single CeO_2 mesoporous sphere. (**d**) Compositional line profile across the single sphere (from A to B) probed by Energy Dispersive X-ray Analysis (EDXA) line scanning. Reproduced from Ref. [74] with the permission from the ACS Publisher.

The catalysts with core-shell and yolk–shell structures also attract attention [75,76]. The $Ag@CeO_2$ catalysts with a core-shell structure were prepared by surfactant-free method with subsequent annealing redox reaction between silver and ceria precursor during co-deposition [77]. The particles with metallic Ag cores with a diameter of 50–100 nm CeO_2 shell with a thickness of 30–50 nm were tested in CO oxidation (catalyst mass was 100 mg, 1% CO, a gas flow of 20 mL/min). The calcination of $Ag@CeO_2$ at 500 °C in air flow led to the growth of catalytic activity (100% CO conversion at ~120 °C) in comparison with freshly deposited precipitate and catalyst after hydrothermal treatment and drying at 80 °C. This growth of activity was attributed to the strengthened interfacial interactions between Ag core and CeO_2 shell during the calcination process (confirmed by TPR-H_2) and to the fast desorption of CO_2 from the surface of catalyst that was shown by Fourier Transform Infrared (FTIR) spectroscopy of adsorbed CO_2. The charge transfer due to enhanced metal–support interaction from Ag to CeO_2 was shown by XPS [39]. It is noteworthy that one-, two- and three-coordinated OH groups were shown to exist over CeO_2 surface [78], and their effect cannot be neglected.

Thus, according to the literature, Ag/CeO$_2$ composites are promising catalysts for CO oxidation. The method of catalyst preparation, shape of ceria nanoparticles, and morphology of ceria are the factors determining the catalytic properties of the composites. Special attention is given to oxygen vacancies, and their concentration depends on the shape of ceria particles, amount of silver and charge states of its clusters/nanoparticles as well as pre-treatment conditions. Certainly, the presence of silver on the surface of ceria promotes the formation of oxygen vacancies and facilitates the growth of catalytic activity in CO oxidation. The features of interfacial interaction also should be considered since the transfer of electronic density from silver NPs to ceria accompanies metal–support interaction in Ag/CeO$_2$ catalysts. These phenomena may play a key role in oxidative catalysis [79,80], reduction of nitroarenes [81], photocatalysis [82]. Different synthetic strategies may be developed to synthesize Ag/CeO$_2$ with high activity in CO oxidation and find real application in industrial or indoor air purification from CO and VOCs.

2.2. Soot Oxidation

Soot is an amorphous impure carbon formed during incomplete combustion of fuels and hydrocarbons in internal combustion engines, coal burning, power-plant boilers, etc. It is formed as a by-product impairing the normal operation of combustion engines by fouling of exhaust systems, generation of exhaust plumes, blocking the pipes, etc. [83]. Soot particles are harmful to the human respiratory system since they cannot be filtered by upper airways. Thus, the development of materials that prevent the harmful impact of soot on the environment and human health is an important research and technology challenge. The soot combustion of diesel exhaust particulate occurs at temperatures above 600 °C, while typical diesel engine exhaust temperatures are in the range of 200–500 °C [84,85]. Therefore, the decreasing of the temperature of soot combustion is the main requirement for catalysts in this reaction.

The contact between soot and catalyst plays a key role in solid–solid reactions, and the observed catalytic activity depends on the gas–solid–solid interaction [86]. The contact conditions between soot and catalyst determine the combustion performance. In the literature two types of catalyst–soot contact studies under laboratory conditions are proposed: tight contact (TC) and loose contact (LC) [85–87]. The LC mode comprises a mixing or shaking of the catalyst–soot mixture with a spatula providing conditions for contact between soot particle and catalyst similar to those over diesel filter. TC mode is achieved by milling (ball or mortar milling) of the mixture during several minutes. Compared to the LC mode, the TC mode is less representative of the real contact conditions but is required to better understand and discriminate the morphologies [86,88].

Many effective catalytic systems have been proposed for soot combustion and other oxidation reactions [83,89]. Due to their unique physical-chemical properties, especially high redox properties and the lability of lattice oxygen, ceria and ceria-based materials also show high catalytic activity in total oxidation reactions, and soot oxidation to carbon dioxide is not an exception. Ceria also possesses high oxygen storage capacity (OSC), which allows using the oxide not only as a support or modifying additive, but also as a catalyst for soot oxidation. A selection of CeO$_2$-based catalysts for soot oxidation is presented in Table 1. In Ref. [90] the catalytic activity of pure ceria prepared by co-precipitation method was described. Precipitation of aqueous solution of HNO$_3$ and Ce(NO$_3$)$_3$ was carried out using the 0.4 M NaOH solution and 0.4 M Na$_2$CO$_3$. Combustion temperature of pure oxide samples was achieved in the region of 445–560 °C. The acidification of cerium precursor at the stage of catalyst preparation improved the catalytic performance of the obtained materials. The sample prepared by precipitation method using HNO$_3$/Ce(NO$_3$)$_3$ = 2 had the highest catalytic activity with T$_m$ = 465 °C. It is noteworthy that the use of large amounts of alkali metals at the stage of synthesis may significantly influence on the morphology and defective structure of cerium oxide, which will impact on the observed catalytic activity [91].

Morphology is known to play an important role in solid–solid reactions, where the number of contact points is a crucial criterion of activity. In Ref. [92] three different morphologies of pure cerium oxide were studied in soot oxidation reaction. The materials comprised (1) ceria nanofibers that capture the soot particles in several contact points, while having low specific surface area (~ 4 m^2/g), (2) solution combustion synthesis ceria having an uncontrolled morphology, but higher specific surface area (31 m^2/g), and (3) three-dimensional self-assembled (SA) ceria stars having high specific surface area (105 m^2/g) and highly available contact points. The latter showed the highest catalytic activity, and the temperature of soot oxidation reduced from 614 to up to 403 °C for TC and to up to 552 °C in case of LC (Figure 4).

Figure 4. FESEM images representing a loose contact mixture of CeO$_2$ SA-stars and soot at × 40,000 (**a**) × 150,000 (**b**) level of magnifications. Reproduced from Ref. [92] with the permission from the Royal society of chemistry.

Comparing to the morphologies in groups 1 and 2, the three-dimensional shape of SA stars may involve more of the soot cake layer that can be a reason for enhancement of the total number of contact points and higher catalytic activity (Figure 5). SA stars also keep their high intrinsic activity after aging.

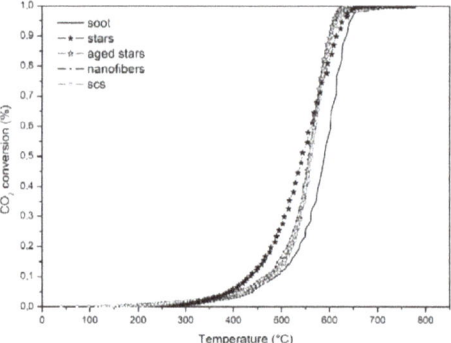

Figure 5. Total soot conversion in loose contact conditions. Reproduced from Ref. [92] with the permission from the Royal society of chemistry.

A comparison of the catalytic performance of pure ceria with different morphology under LC conditions was carried in [60], and the results were compared to those reported in Refs. [46,93–95]. The activity was shown to decrease in the following order: nanorod > nanocube > fiber > flake, and the lowest temperature of complete combustion of 485 °C is observed for nanorod samples.

In Ref. [96] hydrothermal and solvothermal methods were used to prepare nanostructured ceria with different morphology (nanorod, nanoparticle, and flake). The nanorod sample showed the best catalytic activity (soot combustion temperatures for TC and LC modes were 368 and 500 °C, respectively) that was attributed to the maximal amount of adsorbed oxygen species on its surface. Moreover, the high specific surface area, determined by BET (Brunauer Hemmet Teller) method, was pointed out to have a positive effect in improving the activity under the LC mode. In Ref. [97] hydrothermal method was used to prepare conventional polycrystalline ceria and single-crystalline ceria nanorods and nanocubes. The obtained samples differ by the surface formed ((100) surfaces were typical for nanocubes, a mixture of (100), (110) and (111) surfaces for nanorods, while (111) surface was obtained for conventional polycrystalline ceria). More reactive exposed surfaces demonstrated higher catalytic activity and soot oxidation becomes a surface-dependent reaction. Soot, while located at the soot–ceria interface, can reduce ceria, and such surface becomes the source of active superoxide ions. The formation energy of a surface oxygen vacancy is considered important for activity enhancement.

According to Ref. [48], the redox properties of ceria are an important, but not the major factor for catalytic soot oxidation. A comparison of fluorite-type oxides CeO_2, Pr_6O_{11}, CeO_2–ZrO_2, ZrO_2 characterized by high oxygen capacity revealed that the reactivity rather than quantity of oxygen species involved in oxygen release/storage processes is a favorable factor for low-temperature soot oxidation. CeO_2 was shown to be much more active in soot oxidation, than Pr_6O_{11} and CeO_2–ZrO_2 that had higher OSC values than pure CeO_2. Using the electron spin resonance (ESR) method it was demonstrated that the reason was connected with the ability of the CeO_2 surface to generate superoxide ions (O_2^-) that can rapidly react with neighboring carbon or recombine to yield O_2.

Despite unique physical-chemical properties, it is often not feasible to use pure ceria, since a significant loss of specific surface may occur due to thermal sintering, deactivation of redox pair, reduction of OSC leading to deterioration of catalytic activity [98], etc. Even small sintering causes a large impact on the crystallite sizes and the presence of oxygen vacancies, which significantly reduces the catalytic activity. The presence of metal ions in the ceria lattice allows reducing the effects of sintering and loss of catalytic activity along with a significant increase of OSC [99,100].

Special attention should be paid to the effect of introduction of Ag into the CeO_2 structure. Loading of Ag NPs on CeO_2 improves the reactivity of CeO_2 lattice oxygen toward soot oxidation. Kinetic studies showed [45] that lattice oxygen of ceria interacting with Ag NPs had similar reactivity to the one of lattice oxygen in Ag_2O. Ag NPs enhance reducibility of ceria (which was also shown in [101] and was attributed to reverse spillover of oxygen atoms from the Ag–CeO_2 boundary to the Ag NPs along with other possible interpretations), but not the reoxidation ability of reduced ceria surface by dioxygen. Silver can become an agent that allows rapid formation of O_x^-. In Ref. [102] using cyclic H_2-TPR and Raman studies, it was shown that both dissociative adsorption of gaseous oxygen and migration of bulk oxygen of ceria can be facilitated by silver. This results in a rapid generation of atomic oxygen over silver, which under the TC mode can transfer onto soot particle and lead to catalytic oxidation reaction [103]. If not, its spillover onto the ceria surface occurs, and the oxygen transforms to O_x^- through 2O–O_2^-–2O^-–2O^{2-} over the oxygen vacancies [45,49,57,104,105]. On the other hand, silver is proposed to participate in the reverse transformation of O^- to O_2^- [105].

Table 1. A selection of CeO$_2$-based catalysts for soot oxidation.

Catalyst	Preparation Method	CeO$_2$ Morphology	S$_{BET}$, m^2/g	Particle Size, nm Ag	Particle Size, nm CeO$_2$	Ce^{3+}/Ce^{4+} Ratio	Catalyst/Soot Ratio	Contact Mode	Reaction Conditions	T$_{10}$, °C	T$_{50}$, °C	T$_{90}$/T$_{max}$, °C	Ref.
CeO$_2$-NC	hydrothermal	nanocubes	11	-	100	0.45	4:1 (mass.)	TC	1% O$_2$/N$_2$ 500 mL/min, isothermal reactions at 300 °C and 350 °C	-	430	-	[105]
CeO$_2$-NP	thermal decomposition	irregular shaped	71	-	15	0.57				-	458	-	
CeO$_2$-Sp	hydrothermal	spindles	79	-	25	0.53				-	527	-	
CeO$_2$-30	precipitation at 30 °C	irregular shaped	49	-	11	0.52	4:1 (mass.)	LC	20% O$_2$/80% N$_2$	-	-	598	[106]
CeO$_2$-50	precipitation at 50 °C		41	-	15	0.51				-	542	-	
CeO$_2$-70	precipitation at 70 °C		49	-	15	0.50				-	542	-	
Ce-R	hydrothermal	nanorods	80	-	250 nm × 2 μm	Ce^{3+} 25.1 at. %	9:1 (mass.)	LC	10 vol%O$_2$/N$_2$	356	500	554	[96]
								TC		286	368	400	
Ce-P	solvothermal	irregular shaped	88	-	30-40	Ce^{3+} 16.5 at. %		LC		413	521	573	
								TC		320	433	474	
Ce-F	solvothermal	flakes	62	-	25	Ce^{3+} 19.1 at. %		LC		433	554	622	
								TC		306	383	440	
Ce-SAS	hydrothermal route in a batch stirred-tank reactor	SA stars	124	-	10	N/A	45:5 (mass.)	LC	50% air/ 50% N$_2$ constant 100 mL min^{-1}	450	560	610	[107]
								TC		385	415	505	
Ce-NC	hydrothermal	nanocubes	4	-	54	N/A	45:5 (mass.)	LC	50% air/ 50% N$_2$ 100 mL min^{-1}	420	465	575	[93]
								TC		370	385	430	
Ce-ND	thermal decomposition	irregular shaped	72	-	7-35	N/A	45:5 (mass.)	LC		475	530	600	
								TC		360	390	498	
Ce-NC	hydrothermal	nanocubes	4	-	54	Ce^{3+} 27.6 at. %		LC	10% of O$_2$/N$_2$ at rate of 100 cm^3 min^{-1}	417	477	584	[58]
								TC		396	400	425	
Ce-NR	hydrothermal	nanorods	4	-	43	Ce^{3+} 25.5 at. %	45:5 (mass.)	LC		429	536	623	
								TC		381	416	455	
Ce-M	improved grafting	mesoporous	75	-	5	Ce^{3+} 25.5 at. %		LC		398	538	604	
								TC		374	464	510	
Ce-SCS	solution combustion	mesoporous	69	-	35	Ce^{3+} 36.1 at. %		LC		436	580	633	
								TC		392	476	558	
CeO$_2$-CP1-F	co-precipitation	irregular shaped	52.6	-	8.46	Ce^{3+} 21.71 at. %	45:5 (mass.)	LC	5% O$_2$/Ar, 200 mLmin^{-1}	-	-	545	[96]
CeO$_2$-CP2-F	modified co-precipitation HNO$_3$/Ce(NO$_3$)$_3$ = 0.5 (mol)		22.7	-	7.87	Ce^{3+} 12.77 at. %				-	-	530	
CeO$_2$-CP3-F	modified co-precipitation HNO$_3$/Ce(NO$_3$)$_3$ = 1 (mol)		24.6	-	6.05	Ce^{3+} 11.90 at. %				-	-	480	
CeO$_2$-CP4-F	modified co-precipitation HNO$_3$/Ce(NO$_3$)$_3$ = 2 (mol)	irregular shaped	30.13	-	6.07	Ce^{3+} 10.58 at. %	45:5 (mass.)	LC	5% O$_2$/Ar, 200 mLmin^{-1}	-	-	465	[96]
CeO$_2$-CP4-A	CeO$_2$-CP4 calcined at 750 °C for 6 h		1.80	-	47.18	Ce^{3+} 15.60 at. %				-	-	440	
CeO$_2$-S-F	solid combustion	irregular shaped	77.1	-	9.63	Ce^{3+} 26.58 at. %				-	-	540	
CeO$_2$-CA-F	citric acid sol-gel		45.0	-	9.68	Ce^{3+} 30.76 at. %				-	-	560	
CeO$_2$-500	electrospinning with calcination at 500 °C		20.4	-	241-253	N/A	95:5 (mass.)	LC	21% O$_2$ and 79% N$_2$, 100 mL/min	-	-	596	[47]
								TC		-	-	429	
CeO$_2$-800	electrospinning calcination at 800 °C	nanofibers	3.45	-	241-253	N/A		LC		-	-	633	
								TC		-	-	504	
								LC		-	-	639	
CeO$_2$-1000	electrospinning calcinations at 1000 °C		3.40	-	241-253	N/A		TC		-	-	513	
CeO$_2$	precipitation	irregular-shaped	45	-	N/A	N/A	20:1 (mass.)	TC	10% O$_2$/N$_2$ 10 °C min^{-1}	-	-	393	[48]

Table 1. Cont.

Catalyst	Preparation Method	CeO_2 Morphology	S_{BET}, m^2/g	Particle Size, nm Ag	Particle Size, nm CeO_2	Ce^{3+}/Ce^{4+} Ratio	Catalyst/Soot Ratio	Contact Mode	Reaction Conditions	T_{10}, °C	T_{50}, °C	T_{90}/T_{max}, °C	Ref.
CeO_2	precipitation/ripening	nanofibers	4	-	72	N/A	4.5:5 (mass.)	LC	10% O_2/N_2	480	555	560	[92]
								TC		383	439	445	
CeO_2	solution combustion	uncontrolled nanopowders	31	-	45	N/A		LC		483	562	562	
								TC		368	411	417	
CeO_2	hydrothermal	three-dimensional SA stars	105	-	9	N/A	4.5:5 (mass.)	LC	10% O_2/N_2	435	543	552	[62]
								TC		354	410	403	
CeO_2	SA stars aged 5 h at 600 °C	Aged SA stars	50	-	15	N/A		LC		473	559	559	
								TC		381	453	465	
AgCe-NC	incipient wetness impregnation (Ag: 5 wt. %)	nanocubes	10	1.5–3.5	100	0.34	4:1 (mass.)	TC	1% O_2/N_2, 500 mL/min 100,000 h^{-1}, isothermal reactions at 300 °C	-	376	-	[105]
AgCe-NP	incipient wetness impregnation (Ag: 5wt. %)	irregular shaped	64	1.5–3.5	16	0.52				-	389	-	
AgCe-Sp	incipient wetness impregnation (Ag: 5 wt. %)	spindles	69	1.5–3.5	27	0.37				-	411	-	
Ag/CeO_2-30	incipient wetness impregnation (Ag: 5 wt. %)	irregular shaped	37	5	15	0.29	4:1 (mass.)	LC	1% O_2/N_2 after 4 cycles	522	606	691	[104]
Ag/CeO_2-50			33	8	20	0.27				488	596	660	
Ag/CeO_2-70			37	8	15	0.23				504	602	675	
Ag/CeO_2-500	electrospinning calcination at 500 °C (Ag-4.5 wt. %)	nanofibers	5.07	10	241–253	N/A	9.5:5 (mass.)	LC	21% O_2, 79% N_2, 100 mL/min	-	-	481	[17]
								TC		-	-	429	
Ag/CeO_2-800	electrospinning calcination at 800 °C (Ag-4.5 wt. %)		3.07	10	241–253	N/A		LC		-	-	485	
								TC		-	-	484	
Ag/CeO_2-1000	electrospinning calcinations at 1000 °C (Ag-4.5 wt. %)		2.74	10	241–253	N/A		LC		-	-	514	
								TC		-	-	496	
Ag/CeO_2	incipient wetness impregnation (Ag: 10wt. %)	irregular shaped	N/A	N/A	16	N/A	20:1 (mass.)	TC	10% O_2/N_2 10 °C·min^{-1}	-	-	345	[8]
CeO_2–Ag	co-precipitation (Ag: 39 wt. %)	rice-ball	14.7	36	16	N/A	19:1 (mass.)	LC	10% O_2/He at 50 mL/min	-	-	376	[3]
								TC		-	-	315	
Ag(39)/CeO_2	impregnation (Ag: 39 wt. %)		30.1	89	21	N/A		LC		-	-	563	
								TC		-	-	381	
Ag(10)/CeO_2	impregnation (Ag: 10 wt. %)		52.0	60	20	N/A		LC		-	-	526	
								TC		-	-	362	
Ag(3.2)/CeO_2	impregnation (Ag: 3.2 wt. %)	irregular shaped	59.2	28	20	N/A		LC		-	-	550	
								TC		-	-	371	
Ag(1.9)/CeO_2	impregnation (Ag: 1.9 wt. %)		70.0	20	20	N/A		LC		-	-	596	
								TC		-	-	414	
Ag(0.95)/CeO_2	impregnation (Ag: 0.95 wt. %)		78.1	n.d	20	N/A		LC		-	-	610	
								TC		-	-	466	

In Figure 6 an effect of silver loading on the catalytic activity of ceria in soot oxidation is represented. The temperature of soot combustion shifted from 668 °C in case of combustion of pure soot to 393 °C for CeO_2 and to up to 345 °C for the case of Ag/CeO_2 [48].

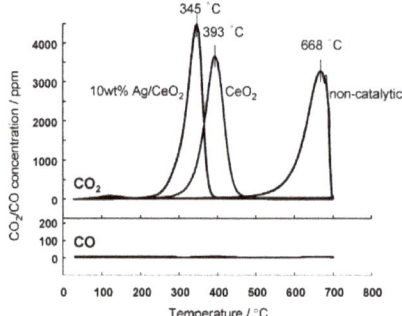

Figure 6. Effect of Ag loading on soot combustion profiles of CeO_2. Soot/CeO_2 tight-contact mixtures with a weight ratio of 1/20 were heated in 10% O_2/N_2 at the rate of 10 °C·min^{-1}. Reproduced from Ref. [48] with the permission from the American chemical society.

By comparing the onset temperature, T_i, of soot oxidation over various metal-loaded CeO_2 with different loadings (Figure 7) it results that Ti can be lowered with an increase of Ag loading from 357 to 324 °C (20 wt %). On the contrary, loading other metals, such as Pd, Pt, and Rh, could not improve the activity. This result supports that superoxides activated over silver are the active species responsible for low-temperature soot oxidation.

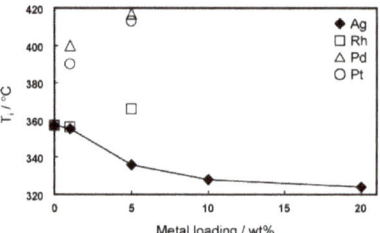

Figure 7. Soot oxidation activity (T_i) of metal-loaded CeO_2 measured in a flow of 10% O_2 and N_2 balance. Tight-contact soot/catalyst mixtures with a weight ratio of 1/20 were heated at the rate of 10 °C·min^{-1}. Reproduced from Ref. [48] with the permission from the American chemical society.

The catalysts for soot combustion have two main drawbacks, i.e. poor soot/catalyst contact and restricted amount of active site. The promising composites should possess relatively low specific surface area and have no micropores and small mesopores, which will provide the presence of a maximal number of active sites on the external surface of the grain and will facilitate the effectiveness of catalyst performance. Various preparation technique can be used to create such active surfaces. While the impregnation method still can be used [48], the relatively simple and economically feasible co-precipitation technique is considered the major way to prepare Ag/CeO_2 catalysts for soot oxidation [44,49,104,108]. As a result, an opportunity exists to design favorable structure to transfer/diffuse the activated oxygen species to reaction zones of the catalyst and promote better catalyst–soot contact.

Among the catalysts prepared by co-precipitation technique, special interest is devoted to those with the "rice-ball" core-shell structure [49,104] comprising metallic Ag particles in the core surrounded by CeO$_2$ particles. These catalysts possess a unique agglomerated structure with a diameter of about 100 nm, where large Ag particles (30–40 nm) and a large interface between the Ag and CeO$_2$ particles cause its excellent catalytic performance in soot oxidation due to this morphological compatibility (the oxidation proceeds below 300 °C).

A less common way to prepare Ag/CeO$_2$ catalysts for soot oxidation is the electrospinning method [47]. CeO$_2$ nanofibers with diameters of 241–253 nm were produced using this method (Figure 8).

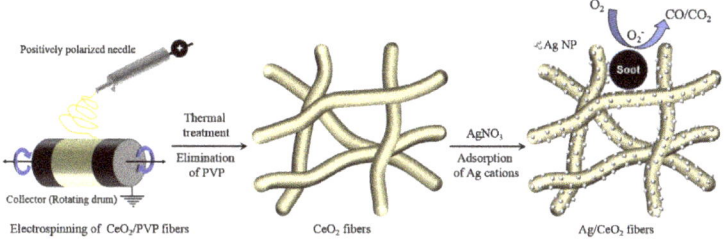

Figure 8. Schematic illustration of Ag/CeO$_2$ nanofiber synthesis sequence. CeO$_2$ nanofibers were fabricated through the electrospinning of spinnable Ce/PVP in a DMF/EtOH precursor solution followed by thermal treatment. Ag was then loaded on the surfaces of the CeO$_2$ nanofibers. Reproduced from Ref. [47] with the permission from the Elsevier.

The Ag/CeO$_2$ and CeO$_2$ fibrous catalysts calcined at 500 °C exhibited an improved catalytic performance in soot oxidation caused by their large pore sizes related to the macroporous characteristics of the porous structure in CeO$_2$. Large surface areas of CeO$_2$ and Ag metallic species can contribute to high soot oxidation activity (Figure 9).

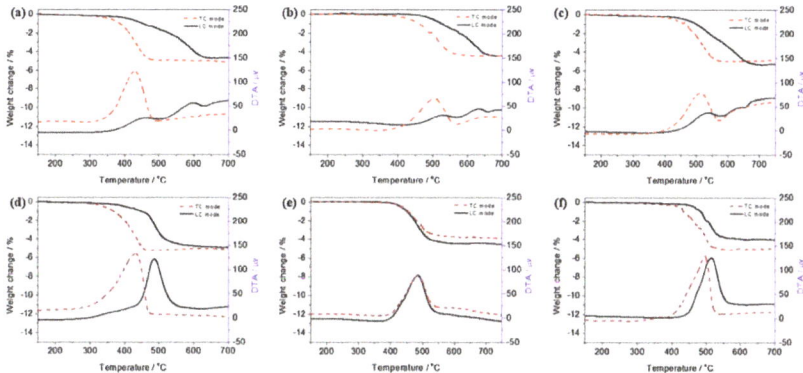

Figure 9. TG and DTG curves: (a) CeO$_2$-500, (b) CeO$_2$-800, (c) CeO$_2$-1000, (d) Ag/CeO$_2$-500, (e) Ag/CeO$_2$-800, and (f) Ag/CeO$_2$-1000. Reproduced from Ref. [47] with the permission from the Elsevier.

In Ref. [106] it is pointed out that under oxygen-rich conditions the activity of Ag/CeO$_2$ catalysts is caused by oxygen vacancies near Ag particles, while under oxygen-poor conditions it is controlled by bulk oxygen vacancies. The generation and transfer of active oxygen are affected by combinations of both types of oxygen vacancies.

The mechanism of soot oxidation over Ag/CeO$_2$ composites is also debating. Soot oxidation is a solid–solid–gas reaction, and there are two points of views on the predominant reaction mechanisms of soot oxidation in the literature [48,109–111]. On one hand, soot oxidation is initiated by the surface-active oxygen (peroxide and superoxide (O^- and $O_2{}^-$) species), which may be activated by the oxygen vacancies. From the other hand, surface active oxygen comes from the bulk by migration of lattice oxygen. Ref. [99,108] describes a mechanism of metal oxide catalyst participation in redox cycle, where metal is subjected to repeated oxidation and reduction according to the following reaction set:

$$M_{red} + O_{gas} \rightarrow M_{oxd}-O_{ads} \quad (1)$$

$$M_{oxd}-O_{ads} + C_f \rightarrow M_{red} + SOC \quad (2)$$

$$SOC \rightarrow CO/CO_2, \quad (3)$$

where M_{red} and $M_{oxd}-O_{ads}$ represent the reduced and oxidized states of the catalyst, respectively; O_{gas} and O_{ads} are gaseous O_2 and surface adsorbed oxygen species, respectively; C_f denotes a carbon active site or free site on the carbon surface, and SOC represents a surface carbon-oxygen complex.

According to this mechanism, atomic O_{ads} species is formed through dissociative adsorption of gas-phase oxygen on the metal oxide surface, and then attacks the reactive free carbon site C_f yielding an oxygen-containing active intermediate. CO/CO_2 are formed through the reaction between the intermediate and either O_{ads} or gas-phase O_2. The authors [45,99,108] suggest that in this mechanism the surface adsorbed oxygen species play the key role in soot oxidation, in contrast to CO oxidation that occurs through the Mars-van Krevelen mechanism. However, some researchers consider that the second reaction mechanism is prevalent in soot oxidation over ceria-based catalysts under real conditions [48] (Figure 10).

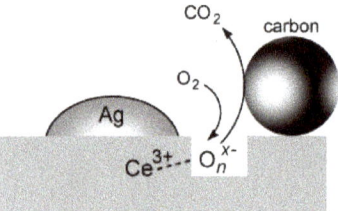

Figure 10. A schematic mechanism of soot oxidation over Ag/CeO_2 catalyst. Reproduced from Ref. [45] with the permission from the Elsevier.

In case of reverse CeO_2–Ag catalyst [49], a synergistic effect of Ag and CeO_2 particles causes adsorption of gas-phase O_2 followed by formation of atomic oxygen species and the process is facilitated due to large Ag–CeO_2 interface. The O species on the silver surface migrates to the surface of ceria particles through the interface and transforms into $O_n{}^{x-}$ species (Figure 11). These atomic oxygen species exist in equilibrium during soot oxidation. Then the mobile active $O_n{}^{x-}$ species migrates onto soot particle through the soot–ceria contact and completely oxidizes the soot into CO_2.

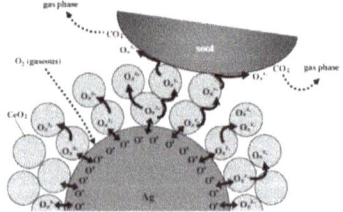

Figure 11. A schematic mechanism for soot oxidation over the CeO_2–Ag catalyst. Reproduced from Ref. [49] with the permission from the Elsevier.

Another important problem that occurs in particulate filters under real conditions is connected with the loss of contact between the catalyst and solid reactant (e.g., unreactive ash). In Ref. [104] the catalytic soot oxidation was shown to occur, when a physical barrier of ash deposit exists between the catalyst and the solid soot, and the reaction proceeds without a direct catalyst–soot contact or any external energy applied (Figure 12). A CeO$_2$–Ag catalyst prepared by the co-precipitation and a Ag/CeO$_2$ catalyst prepared by impregnation showed catalytic activity for remote oxidation of soot separated by the deposition of alumina or calcium sulfate, while CeO$_2$ catalyst did not. The remote oxidation effect is extended to more than 50 µm for both the CeO$_2$–Ag and Ag/CeO$_2$ catalysts, with the highest effect over the former catalyst. Based on the results of the ESR experiments, a mechanism for the observed phenomenon was proposed, in which a superoxide ion (O$_2^-$) generated on the catalyst surface first migrated to the ash surface and then to the soot particles and then subsequently oxidizes it.

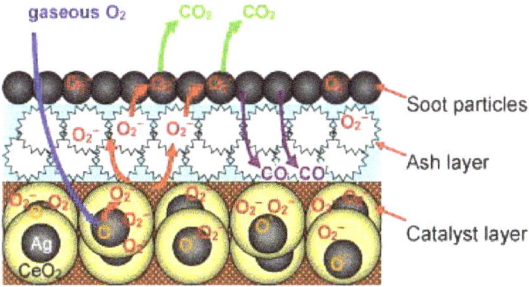

Figure 12. A schematic mechanism for remote catalytic soot oxidation over a catalyst composed of Ag and CeO$_2$. Reproduced from Ref. [104] with the permission from the Elsevier.

In [112] several model Ag/CeO$_2$ catalysts with uniform structures and diverse surface oxygen vacancy (V$_{O-s}$) contents were prepared by solution combustion method, and the processes of their activation and deactivation were considered (Figure 13). The V$_{O-s}$ content, conditions of catalyst–soot contact and extra oxygen supplier were pointed out as the most important structural factors in the activity of soot oxidation catalysts. The dioxygen concentration in the reaction atmosphere was assumed to influence the V$_{O-s}$ content, while ceria reduction was mentioned to occur around the catalyst–soot contact points and did not take place in the presence of O$_2$. Moderate amounts of V$_{O-s}$ were shown to boost the catalytic activity by generating more O$_x^{n-}$ species, while their excess yields O^{2-} instead of O$_2^-$ that hinders the process. The interfacial reduction of ceria and insufficient O$_2^-$ delivery and regeneration were suggested to determine the catalyst performance. The deactivation can be postponed by noble metal addition, resulting in accelerated soot combustion over noble metal-containing catalysts.

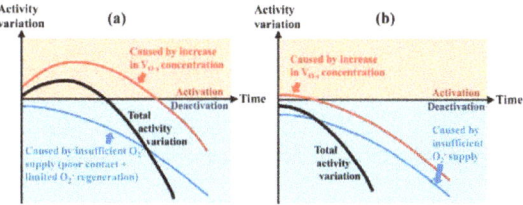

Figure 13. Schematic explanation for activity variations over (**a**) Ag/CeO$_2$ with low initial V$_{O-s}$ concentration (e.g., AgCe-0, AgCe-0.01, AgCe-0.02 and AgCe-0.03) and (**b**) Ag/CeO$_2$ with high initial V$_{O-s}$ concentration (e.g., AgCe-0.04 and AgCe-0.05) during isothermal soot oxidation. Reproduced from Ref. [112] with the permission from the Royal Society of Chemistry.

Thus, the development of catalysts with a special state of the deposited phase, characterized by a strong metal–support interaction, makes it possible to stop the migration of the deposited particles of the active phase, preventing the process of thermal aging (sintering) of the catalyst, which is one of the main problems in the operation of catalytic systems for cleaning emissions of internal combustion engines, both gasoline and diesel. The synergistic effect of Ag/CeO$_2$ catalysts is determined by high activity, stability and is achieved by decreasing the costs for use of expensive metals, e.g., platinum [113–115], with saving of efficiency in the processes of catalytic cleaning of emissions of internal combustion engines. In this way, Ag/CeO$_2$ composites are considered promising catalysts for soot oxidation.

2.3. VOCs Abatement

VOCs are a large group of organic chemicals having high vapor pressure and low boiling point at atmospheric pressure (these include, but are not limited to aldehydes, alcohols, aromatic compounds, etc.). These properties cause evaporation or sublimation of these compounds from liquid or solid state and entering the indoor and outdoor air. VOCs are known to possess high toxicity, poison the atmosphere and have a negative impact on human health and the environment [34]. To date, numerous ways to solve the challenge of air pollution, such as combustion of wastes, biodegradation [116], adsorption [117], plasmochemical decomposition [118], photocatalytic oxidation [119], ozonation [120], etc., have been proposed. The main drawback of these methods is the high-energy consumption that may be accompanied by the formation of formaldehyde and CO as well as the complexity of regeneration of the active phase (bacteria, adsorbents, photocatalysts). Catalytic oxidation of VOCs to carbon dioxide and water are considered the most promising methods to control the emissions [121–124]. The use of catalysts allows carrying out VOCs oxidation at relatively low temperatures at complete conversion. As a rule, two main types of effective catalysts for total oxidation of VOCs are developed, including supported metals (e.g., Au, Pt, Pd, Ag) [125–130] and transition metal oxides (CeO$_2$, MnO$_2$, Co$_3$O$_4$) [130–133]. The combination of noble metal and transition metal oxide used as a support or modifier is promising to increase the effectiveness of catalytic composites [134–136].

Currently, Ag–CeO$_2$ composites represent both scientific and practical interest as catalysts for VOCs abatement, in particular oxidation of formaldehyde, methanol, toluene, acetone, etc. A selection of literature data on Ag/CeO$_2$ composites used in VOCs abatement is presented in Table 2. Several articles were published on formaldehyde oxidation over Ag/CeO$_2$ catalysts [40,137–139]. One of the pioneer works in this field was carried out by S. Imamura et al. [140], who suggested using Ag/CeO$_2$ as catalysts for formaldehyde oxidation. High activity of the Ag/CeO$_2$ composite was suggested to be governed by high dispersion of active silver on CeO$_2$ and easier removal of surface oxygen as compared to the one over individual Ag or CeO$_2$ components. The authors pointed out that compared to other group VIII metals, silver is less expensive and more abundant and shows high activity and durability, when high temperatures are not required.

Table 2. Literature data on catalytic VOCs abatement over Ag/CeO$_2$ composite catalysts.

Type of VOC	Preparation Method	Loading of Ag, wt. %	T$_{VOC\ conv.}$, °C	S, m^2/g	Mean Ag NP Diameter (nm)	Reaction Conditions	TOF × 10^3, s^{-1}	T, °C	Ref.
CH$_2$O	CP	61.3 28.4 15 7.69	80%: 150	40.5 to 84.4	N/A	1 mL of catalyst, CH$_2$O: 0.42%, methanol: 0.074%, H$_2$O: 19.9%, N$_2$: 62.7%, O$_2$: 16.9% GHSV = 21,000 h^{-1} T$_{range}$: 423–573 K	-	-	[140]
CH$_2$O	WI	8	100%: 125	113.7	N/A	110 ppm of CH$_2$O 20% O$_2$, N$_2$ balance GHSV = 100,000 mL (g$_{cat}$·h)$^{-1}$ Kinetic studies: 1400 ppm of CH$_2$O. GHSV = 302,000 mL (g·cat·h)$^{-1}$	6.8	100	[137]
CH$_2$O	WI	1	100%: 100	70.8	<3	50 mg of catalyst 600 ppm CH$_2$O 20.0 vol% O$_2$, N$_2$ balance GHSV = 120,000–360,000 h^{-1}	1.8	100	[18]
CH$_2$O	HT WI	2	100%: 110	HT: 125.4. WI: 55.5	nanospheres HT: 14.8 WI: 2.4	50 mg of catalyst powder mixed with quartz sand 810 ppm of CH$_2$O 20% O$_2$, N$_2$ balance GHSV = 84,000 h^{-1}	5.0	110	[40]
CH$_2$O	HT WI	5	100%: 110 (nr) 50%: nr: 74 np: 89 nc: 108	nr: 128.46 np: 104.74 nc: 72.63	np: 4.0 nr: 6.0 ± 2.0 nm and 50.0–100.0 nm	50 mg of catalyst 810 ppm of CH$_2$O GHSV = 84,000 h^{-1} contact time was 0.34 s T$_{range}$: 30–240	1.9* TOF$_{Ag}$ nr: 71.0 np: 46.0 nc: 31.0	100	[139]
propylene	WI DP	10	50%: WI: 173 DP: 261	WI:92 DP:84	N/A	100 mg of fine catalyst powder, air and 6000 ppm of C$_3$H$_6$, reactive flow of 100 mL·min^{-1}	WI: 2.2 DP: 0.13	170	[41]
propylene	WI DPU IRC	4	50%: WI: 221 DPU: 260 IRC: 200	WI: 149 DPU: 123 IRC: 99	N/A	200 mg of catalyst 6000 ppm of C$_3$H$_6$ a total flow of 100 mL/min T$_{range}$: 60–400	WI: 0.27 DPU: 0.22 IRC: 0.14	170	[141]
propylene	WI DP	2.14	50%: WI: 220 DP: 245	WI: 98 DP: 118	N/A	100 mg of fine catalyst powder 6000 ppm of C$_3$H$_6$ a total flow of 100 mL·min^{-1} T$_{range}$: 100–400	WI: 0.8 DP: 0.5	170	[142]
toluene			50%: WI: 240			2000 ppm of C$_7$H$_8$	0.34	170	

Table 2. Cont.

Type of VOC	Preparation Method	Loading of Ag, wt. %	$T_{VOC\ conv.}$, °C	S, m^2/g	Mean Ag NP Diameter (nm)	Reaction Conditions	TOF × 10^3, s^{-1}	T, °C	Ref.
toluene		4.8	50%: DP: 265 CP: 260			0.7 vol.% VOC 10 vol.% O_2 He balance GHSV = 7.6 × 10^{-3} $molVOC\ h^{-1}\ gcat^{-1}$	-	-	
methanol	DP CP	4.7	50%: DP: 131 CP: 113	DP: 112 CP: 130	DP: 7.1–6.7 CP: <4–3.3		-	-	[54]
acetone			50%: DP: 225 CP: 220				-	-	
naphthalene	WI	1	100%: 240 50%: 175 (1 wt.% Ag)	143	8.7	120 ppm naphthalene 10% O_2, N_2 balance total gas flow rate was 400 mL/min GHSV = 175,000 h^{-1} T_{range}: 160–300	1.5	170	[143]

WI—wetness impregnation, DP—deposition–precipitation, DPU—deposition–precipitation with urea, IRC—impregnation–reduction with citrate, HT—hydrothermal synthesis, nr—nanorod, np—nanoparticle, nc—nanocube. *—the calculation was carried out as a ratio of mole of converted formaldehyde per mole of Ag loading in the catalysts.

Thus, in Refs. [137,138] a comparison of Ag/CeO$_2$ catalysts with Ag-containing catalysts supported on various supports and the those with different active components supported on CeO$_2$ was considered. Catalytic activity toward formaldehyde oxidation was shown to strongly depend on the Ag particle size and dispersion and the amount of active oxygen species [137]. The 100% formaldehyde conversion was achieved above 125 °C. In Ref. [138] the defective sites of mesostructured CeO$_2$ support prepared by pyrolysis of oxalate precursor were suggested to increase oxygen vacancies able to absorb and activate dioxygen, and highly dispersed silver particles promote this process. This allowed achieving the complete formaldehyde conversion at 100 °C and was accompanied by a strong synergistic interaction between active component and CeO$_2$ support causing enhancement of redox capability of the catalyst.

L. Ma et al. [40] also pointed out the synergistic interaction between Ag and CeO$_2$ that caused an activity enhancement of Ag/CeO$_2$ nanosphere catalysts with average sizes around 80–100 nm composed of small particles with a crystallite size of 2–5 nm as compared to normal Ag/CeO$_2$ particle catalysts prepared by conventional impregnation method. The complete formaldehyde conversion was achieved above 110 °C, which was also explained by the fact that surface chemisorbed oxygen can be easily formed on the Ag/CeO$_2$ nanosphere catalysts. Silver facilitated oxygen activation, which was considered an important aspect of formaldehyde oxidation.

Similar idea was reported in Ref. [139], where the comparison of catalytic properties of Ag/CeO$_2$ catalyst with different morphologies (nanorods, nanoparticles, and nanocubes) of CeO$_2$ prepared by hydrothermal and impregnation method was carried out. The authors pointed out shape dependence of the chemical state of ceria-supported Ag NPs, with the catalysts supported on CeO$_2$ nanorods showing the highest activity caused by the highest surface oxygen vacancy concentration, high low-temperature reducibility as well as existence of lattice oxygen species and lattice defects formed with the participation of both silver and ceria. The electronic silver–ceria interaction yielded Ag^0 in Ag/CeO$_2$ composites, and the Ag^0/(Ag^0 +Ag$^+$) ratio was found the highest for the catalysts supported on ceria nanorods. These results show that the catalytic activity of Ag/CeO$_2$ composites toward formaldehyde abatement can be regulated by engineering the proper shapes of CeO$_2$ supports.

One of the main parameters that allows comparing the catalytic activity of different materials is a TOF. Table 2 presents the TOF values calculated by the authors. Unfortunately, the differences in calculation methods and absence of required experimental information in original papers did not allow comparing the activity of Ag/CeO$_2$ materials correctly.

Besides formaldehyde, Ag/CeO$_2$ catalysts were also used to oxidize other VOCs, e.g. methanol, toluene, acetone, and naphthalene [41,54,141–143]. In these articles, a comparison of catalysts prepared by different methods was represented. The authors attempted to determine the influence of the preparation method and structure of the catalyst on its catalytic activity. Thus, in Ref. [54] the properties of catalysts prepared by deposition–precipitation and co-precipitation methods were compared in total oxidation of methanol, acetone, and toluene. The catalysts prepared by co-precipitation method were revealed to be more active in oxidation reactions. Small crystallites of silver and ceria enhanced the mobility and reactivity of oxygen species over ceria surface, which participated in the said reactions through the Mars-van Krevelen mechanism. The reactivity of the VOCs changed in a row: methanol > acetone > toluene.

In Refs. [41,142] the comparison of the catalytic activity of M/CeO$_2$ composites (M = Au, Cu, Ag) prepared by conventional wet impregnation and deposition–precipitation methods was carried out in propylene oxidation. It was shown that the Ag-containing catalyst prepared by conventional wet impregnation method possessed higher catalytic activity. In Ref. [142] the presence of silver in high oxidation state was considered responsible for high catalytic activity of Ag/CeO$_2$ composites. Using EPR technique it was shown that this is connected with the presence of Ag^{2+} ions (isotopes ^{107}Ag^{2+} and ^{109}Ag^{2+} were detected) along with Ag$^+$ and Ag^0 in the Ag/CeO$_2$-Imp sample, while this was not observed in case of Ag/CeO$_2$-DP (Figure 14, A) [41].

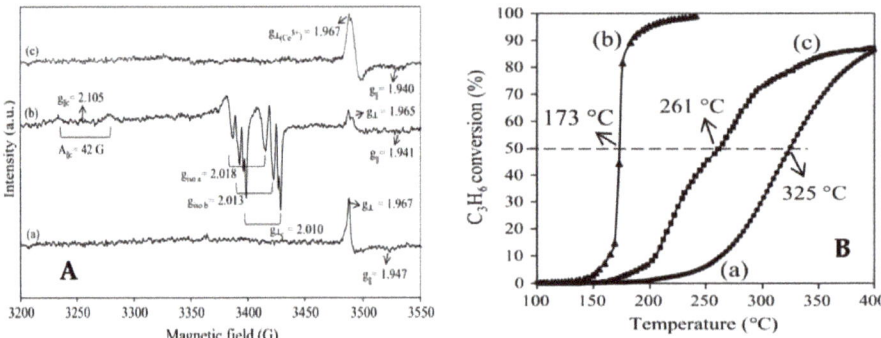

Figure 14. (**A**) EPR spectra for (a) CeO$_2$, (b) 10% Ag/CeO$_2$ (Imp), and (c) 10% Ag/CeO$_2$ (DP) recorded at −196 °C after treatment under vacuum for 30 min. (**B**) Propylene conversion over (a) CeO$_2$, (b) 10% Ag/CeO$_2$ (Imp), and (c) 10% Ag/CeO$_2$ (DP). Reprinted from [41] with the permission of the Elsevier.

In the presence of Ag^{2+} ions, a mobility of some oxygen species increases, which sets conditions for the formation of three redox couples (Ag^{2+}/Ag$^+$, Ag^{2+}/Ag^0, and Ag$^+$/Ag^0). Nitrate precursor decomposition with the participation of O^{2-} of ceria lattice was considered a source of Ag^{2+} ions, while the regeneration of oxygen vacancy may occur either from nitrate or from gaseous oxygen:

$$(Ag^+ + NO_3^-)/(O^{2-}Ce^{4+}O^{2-}) \rightarrow$$

$$(Ag^+ + NO_2^-)/(O^{2-}Ce^{4+}O^{2-}) \rightarrow$$

$$(Ag^{2+} + O^{2-})/(O^{2-}Ce^{4+}O^{2-}) + NO_2$$

In Figure 14B the catalytic conversion of propylene over CeO$_2$, Ag/CeO$_2$-Imp and Ag/CeO$_2$-DP is shown. Adding Ag to CeO$_2$ enhanced the catalytic activity, moreover, the performance of the Imp catalyst was better than that for the DP. In order to evaluate the stability of the catalyst over time, the authors also presented both static (isothermic conditions at 175 °C) and dynamic (7 consecutive cycles vs temperature in the range from 50 to up to 300 °C) aging tests for the activity of the 10% Ag/CeO$_2$ (Imp) sample in propene oxidation. Moreover, EPR studies were carried out for the samples before and after catalysis. It was stated that after catalysis the Ag^{2+} ions retained on the ceria surface. This allows formulating the key role of Ag^{2+}/Ag$^+$ and Ag^{2+}/Ag^0 redox couples as active species in propene oxidation over 10% Ag/CeO$_2$ by prepared impregnation method.

S. Benaissa et al. [141] prepared a mesoporous CeO$_2$ using nanocasting pathway with SBA-15 as a structural template and cerium nitrate as a CeO$_2$ precursor and compared the properties of catalysts on the basis thereof prepared by wetness impregnation (WI), deposition–precipitation with urea (DPU) and impregnation–reduction with citrate (IRC) methods, with the latter being the most active and stable (the catalytic activity and selectivity did not significantly change after 50 h). The authors connected this with higher surface lattice oxygen mobility over this catalyst and with strong silver–mesoporous ceria interaction.

The authors [143] carried out isothermal naphthalene oxidation comparing the activity of catalysts with different Ag content (0.5–5 wt. %), with the sample containing 1 wt. % Ag being the most active one. This was explained by the balance between two factors: oxygen availability and oxygen regeneration capacity. Introduction of Ag to CeO$_2$ was shown to increase both factors. Regeneration capacity was related to the number of oxygen vacancies in bulk ceria, and Ag facilitated the process by reverse spillover effect. Ce^{x+} ions were suggested to be the main active sites. Impregnated silver was claimed to serve as a "pump" and increase bulk oxygen vacancies, while reducing the surface ones, which resulted in oxygen availability and determined the oxygen regeneration. Spillover effect was proposed to reduce the regeneration ability of active oxygen, when Ag loading is high, which was connected with lower concentration of surface oxygen vacancies.

Of particular interest is the approach to locate the Ag/CeO$_2$ composition on the inert support, which is usually represented by alumina or silica [144,145]. Thus, H. Yang et al. [144] used 3DOM CeO$_2$–Al$_2$O$_3$ as a support for Ag catalysts for toluene oxidation. This support was prepared using the Pluronic F127 (EO$_{106}$PO$_{70}$EO$_{106}$) and PMMA as soft and hard templates, respectively. The obtained support showed high-quality 3DOM architecture with a diameter of macropores of 180–200 nm, where ordered mesopores with a diameter of 4–6 nm were formed on the skeletons of macropores. Such structure allowed producing the particles of active component with sizes of 3–4 nm that were evenly distributed on the catalyst surface. The 50% and 90% toluene conversion (1000 ppm) over 0.81Ag/3DOM 26.9CeO$_2$–Al$_2$O$_3$ sample was achieved at 308 and 338 °C, respectively.

In Ref. [145] silica gel prepared by sol–gel method and subjected to hydrothermal treatment was used as a primary support. Ceria and then silver were supported onto silica gel using consecutive impregnation method. The activity of the obtained catalysts was studied in formaldehyde oxidation reaction. The author pointed out that the activity of Ag/CeO$_2$/SiO$_2$ catalysts was significantly higher than the one of Ag/SiO$_2$ sample, which was attributed to synergetic action between silver and ceria. The results obtained for the silver catalyst with small amounts of ceria were not significantly inferior to silver supported over bulk ceria (Figure 15). Thus, the silica-supported ceria-modified silver catalyst can be used for formaldehyde oxidation.

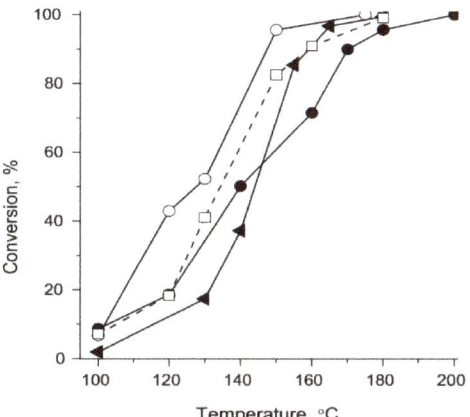

Figure 15. Temperature dependence of the formaldehyde conversion over the catalysts: ●, Ag/SiO$_2$; ○, Ag/CeO$_2$/SiO$_2$; ◀, Ag/MnO$_x$/SiO$_2$; □, Ag/CeO$_2$–MnO$_x$/SiO$_2$. Reaction conditions: 18,000–22,000 ppm of CH$_2$O in dry air; catalyst mass 145 mg; WHSV = 69,000 mL/(gcath). Reproduced from [145] with permission from Elsevier.

To conclude, Ag/CeO$_2$ catalysts are promising materials for VOCs abatement. Even though their activity is inferior to the one of catalysts based on noble metals, their use still represents wide interest due to lower costs. Moreover, the opportunity to increase their activity due to the application of various preparation methods as well as changing of Ag/Ce ratio forms the ground for future research in this field. It is noteworthy that in the literature there is no consensus on the effect of preparation method of Ag/CeO$_2$ composites on their catalytic activity in VOCs abatement.

3. Ag/CeO$_2$ Composites: Insights from Theory

Due to low amounts of silver that are usually used in the preparation of highly effective Ag/CeO$_2$ composites for total oxidation of VOCs, soot, and CO, not all experimental techniques can provide a representation of silver–ceria interface and the ways it works in the said catalytic transformations. Thus, Ag/CeO$_2$ composites have attracted the attention of theoretical chemists. Two main directions are considered: (1) adequate representation and modeling of regular and defective ceria surfaces [132,146–151], (2) systematic studies of the adsorption behavior of Ag clusters on ceria surfaces [152–160]. In the latter case, the structure of Ag–ceria interface is widely discussed, while the adsorption behavior of adsorbates over such composites and their roles in tuning the interfacial properties are modeled in a lesser extent [152].

Researchers point out several difficulties in terms of theoretical modeling of CeO$_2$-based composites. These difficulties are as follows: (1) density functional theory (DFT) does not predict correctly the localized nature of Ce 4f states, (2) change of Ce oxidation state causes incorrect lattice parameters, (3) the calculation results strongly depend on the used methods and functionals, and the obtained energy values oscillate.

These issues were partially addressed by application of hybrid functionals [132,161,162] or DFT+U approach [152,157,163]. The latter is connected with the inclusion of U term for highly correlated Ce 4f electrons in reduced ceria providing partial occupancy of the corresponding atomic level and increasing the accuracy of modeling of the on-site Coulomb interactions in CeO$_2$-based materials. The values for U are usually selected semiempirically. The formalism by Dudarev et al. [164] is usually used. A combination of local density approximation (LDA) and generalized gradient approximation (GGA) in periodic calculations is shown to adequately describe geometry and energy parameters [165] under this approach. However, it is noteworthy that the results of DFT+U calculations depend on many parameters (e.g., lattice constants), which requires special attention to their interpretation.

In Ref. [160] using LDA+U and GGA+U DFT approaches with different U values and periodic slab surface models, charge transfer was shown to occur from Ag to ceria with a concomitant reduction of one Ce surface atom of the top layer, and the transferred electron was localized on Ce atoms. For Ag-based systems, the most favorable adsorption site comprised three surface oxygen atoms. In Ref. [159] the studies of surface structures and electrophilic states of Ag adsorbed on CeO$_2$(111) revealed that charge redistribution can be caused by local structural distortion effects. The distribution of charge was not uniform over the top O layer because of Ag clusters on the underlying O ions, which increased the ionic charge of the remaining O ions and decreased the effective cationic charge over Ce atoms bonded with uncovered O atoms. This also influenced back on the structure of Ag cluster. Silver clusters were shown to induce changes in the oxidation state of several Ce atoms located in the top layer (Ce^{4+} to Ce^{3+}), which are accompanied by a charge flow from metal cluster to surface caused by electronegativity difference between Ag and O atoms [154].

In Ref. [158] charge redistribution during Ag adsorption was confirmed by construction of spin density isosurfaces and site projected density of states. The distortions of selected Ce–O distances were imposed to study the energetics of Ce^{4+} to Ce^{3+} reduction. Oxidation of Ag^0 to Ag$^+$ was assumed, while the probable formation of partially oxidized Ag$_x$O$_y$ species was not considered. Two nearest neighbor Ce^{3+} sites relative to Ag showed the highest Ag adsorption energy at O bridge sites, while three nearest neighbor Ce^{3+} sites showed the highest Ag adsorption at Ce bridge sites.

DFT calculations were carried out for ceria-supported 4-atom transition metal (including Ag) clusters in Ref. [155] and showed that the strength of metal–metal and metal–oxygen interactions depended on the hybridization of d-states of metal with p-states of oxygen as well as the occupation of antibonding Ag d-states. The interactions changed the itinerant f-states of cerium to localized ones, which created a lateral tensile strain in the top layer of Ce on the surface. It was suggested also that the structure of Ag cluster determined the number of cerium atoms in the localized Ce^{3+} oxidation state.

Combined experimental (XPS, STM) and theoretical (DFT+U) approaches were used to study the nucleation and growth of Ag nanoparticles deposited on stoichiometric and reduced thin CeO$_2$

films grown on Pt(111) [157]. A direct electron transfer from Ag clusters and nanoparticles to ceria was reported, and its extent, as well as spin, localization depended on the level of theory used. Ag atoms or nanoparticles supported on stoichiometric CeO_2 acted as electron donors and are subjected to spontaneous direct oxidation at the expense of ceria followed by reduction of Ce ions of the support. The energy costs to move single O atom from ceria toward adsorbed Ag nanoparticle was high, and reverse spillover of oxygen cannot be considered a favorable mechanism of ceria reduction.

Silver–ceria interaction is often compared with the one in Au/CeO_2 and Cu/CeO_2 systems. Due to relatively lower ionization potential, Ag and Cu show higher adsorption energies. Moreover, silver nanoparticles act as a platform for oxygen diffusion leading to partially oxidized Ag nanoparticles located on the surface of the partially reduced ceria [157]. To quantitatively explore the interactions between silver and ceria, a method is proposed utilizing the conversion of total adsorption energy into the interaction energy per Ag–O bond and measurement of a deviation of Ag–O–Ce bond angle from the angle of the sp^3 orbital hybridization of O atom [153]. It is noteworthy that coordination number of O atom, although generally considered, is not included into the correlation, while in Ref. [156] multiple adsorption configurations are shown to exist over single adsorption sites for $Ag/CeO_2(100)$, and electron charge transfer occurs between the neutral silver atom and neighboring Ce^{4+} cation.

In Ref. [152] the reactivity of Ag-modified $CeO_2(111)$ surface used in soot combustion was considered. The interactions of stoichiometric and reduced CeO_2 (111) surfaces with dioxygen, carbon clusters, isolated Ag atoms and silver clusters were studied using DFT+U approach. Carbonaceous species yielded oxygenated carbon moieties of reduced ceria. Peroxo and superoxo species are shown to form, when O_2 is adsorbed over Ag cluster. The role of Ag atoms is to act as a donor, which, when oxidized, donate the valence electron to ceria yielding reduced Ce^{3+} ions. The presence of small Ag clusters mediates the formation of oxygen vacancies (Figure 16).

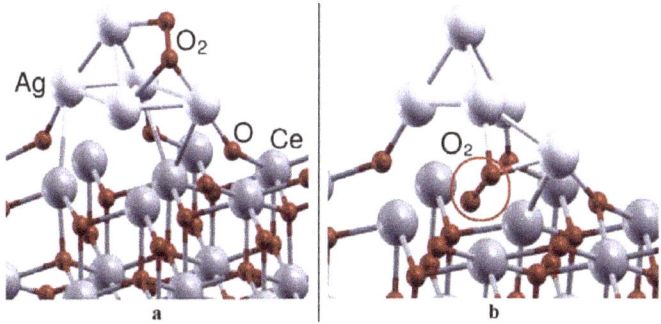

Figure 16. Structure of O_2 adsorbed on $Ag_5/CeO_{2-x}(111)$ surface complex. (**a**) Isomer where O_2 is above the Ag cluster and forms a superoxo species (less stable); (**b**) isomer where O_2 is below the Ag cluster and forms a peroxo group (more stable). Reproduced from Ref. [152]. Copyright© 2011, Elsevier.

The vacancies possess stronger affinity with respect to oxygen as compared to silver that leads to refilling of the cavities with dioxygen. Co-presence of Ag clusters and reduced ceria lightens electron transfer and activation of dioxygen molecule. Silver atoms perform as alkali metal promoters to facilitate O_2 to O_2^- transition that leads to the formation of reduced Ce^{3+} ions. However, partial oxidation of silver can take place in this case.

Despite thorough investigations, still there are several debating issues in the theoretical description of Ag/CeO_2 composites. Among them are the mechanism of oxygen replenishing in the support, different behavior of CeO_2 surfaces, adsorption of silver atoms over long and short O–O bridge sites, quantitative description of Ag–CeO_2 interactions, etc.

4. Emerging Applications

4.1. Photocatalysis

The wide application of CeO_2-based catalysts in oxidative catalysis is mainly attributed to intrinsic redox properties [166]. Conversely, the interest in using ceria in photocatalysis is much lower. This is connected with fast recombination of photoinduced electron–hole pairs and limited visible light adsorption capacity [167]. CeO_2 is an n-type semiconductor with a relatively wide bandgap (E_g = 3.15–3.2 eV) [167,168]. On the other hand, CeO_2 has emerged as a promising material for photocatalysis owing to its chemical stability and photocorrosion resistance [169]. Redox $Ce^{4+} \leftrightarrow Ce^{3+}$ transition is accompanied by oxygen vacancy formation, which has high importance for both oxidative catalysis and electron–hole separation/recombination in photocatalyst [170]. Thus, in Ref. [171] a mesoporous nanorod-like ceria prepared by microwave-assisted hydrolysis of $Ce(NO_3)_3 \cdot 6H_2O$ in the presence of urea was characterized by significant shifts of adsorption to the visible region (a band gap of 2.75 eV) that was associated with the presence of Ce^{3+}. The growth of temperature was also shown to result in significant reduction of the recombination of photogenerated electron–hole pairs. The increased photocatalytic activity in gas-phase oxidation of benzene, hexane, and acetone was found for the prepared mesoporous nanorod-like ceria due to these two phenomena. Thus, the shape of ceria nanoparticles and the presence of Ce^{3+} in the structure provided a growth of photocatalytic activity, including the one under visible light.

Various strategies are being developed to improve the photocatalytic properties of ceria-based materials: morphology control [172,173], doping by europium or yttrium [174,175], fabrication of heterojunctions [176], etc. Thus, in Ref. [172] the degradation of the azo dye acid orange 7 (AO7) under ultraviolet irradiation over hierarchical rose-flower-like CeO_2 nanostructures (Figure 17) is studied. The synthesis of CeO_2 sheets active under the visible light is described in Ref. [173].

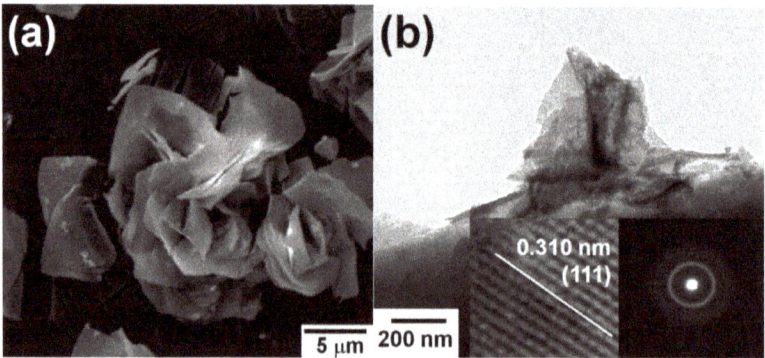

Figure 17. (a) Scanning electron microscopy image and (b) TEM image of the CeO_2 nanopetaled rose-flower-like morphology annealed at 300 °C for 3 h. Insets present a high-resolution TEM image and a selected-area electron diffraction pattern of the CeO_2 roses. Reproduced from Ref. [172] with the permission from the American Institute of Physics.

Moreover, the fabrication of CeO_2-based heterostructures is a more promising way to reduce the band gap and provide improved electron–hole separation due to charge transfer through the interfacial boundaries. Silver salts may be used in photocatalysis due to their semiconductors properties. Thus, Ag_3PO_4 are characterized by relatively small band gap (2.36–2.43 eV) [177], absorb visible light (has yellow color) and possess a good photocatalytic stability. In Ref. [178] the photocatalytic activity of new composite Ag_3PO_4/CeO_2 in degradation of methylene blue and phenol under visible light and UV light irradiation was studied. The photocatalytic activity of the Ag_3PO_4/CeO_2 composite was shown to be associated with the fast transfer and efficient separation of electron–hole pairs at the interfaces of two semiconductors (CeO$_2$ and Ag$_3$PO$_4$). The stability of photocatalyst was demonstrated during five catalytic cycles.

The photocatalytic remediation of water polluted by some chemically stable azo dyes using Ag$_2$CO$_3$/CeO$_2$ microcomposite under visible light irradiation was studied in Ref. [179]. The enhanced photocatalytic activity for the photodegradation of enrofloxacin in aqueous solutions over Ag$_2$O/CeO$_2$ composites under visible light irradiation was demonstrated in Ref. [167]. The composite was synthesized by an in situ loading of Ag$_2$CO$_3$ on CeO$_2$ followed by thermal decomposition. The p-n heterojuction between two semiconductors provided efficient separation of photoinduced charges through the contact of semiconductors that was shown by photoluminescence spectra (Figure 18a). The formation of Ag nanoparticles was associated with photoreduction of Ag$_2$O. The surface plasmon resonance (SPR) on Ag NPs may lead to the formation of electrons and holes in such a way that the electrons could migrate from Ag NPs to the conduction band (CB) of Ag$_2$O (Figure 18b). Thus, Ag NPs may play a specific role in photocatalytic degradation of organic pollutants.

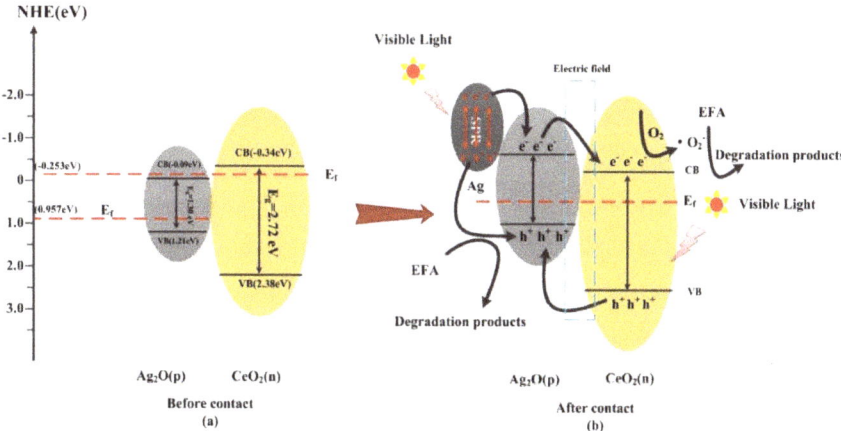

Figure 18. The proposed mechanism for the enhancement of photocatalytic activity of Ag$_2$O/CeO$_2$ catalyst in degradation of enrofloxacin. Reproduced from Ref. [167] with the permission from the Elsevier.

The same effect of photoreduction of silver compounds with the formation of Ag NPs was observed for Ag/AgCl–CeO$_2$ catalysts [180]. The energy of hot electrons, generated on Ag NPs due to SPR, is between 1.0 and 4.0 eV [181], and these electrons could migrate to the CB of AgCl in such a way that the electrons and holes generated on CeO$_2$ and Ag NPs would be efficiently separated. Thus, in composite photocatalysts the role of Ag NPs in visible light adsorption and separation of charges is high.

The decoration of ceria by metals (Au, Pt, Pd, Ag) provides growth of photocatalitic activity due to increased electron–hole separation and extended time of light response of semiconductors [170]. The three main phenomena of charge transfer are involved through metal–semiconductor interface: Schottky barrier (transfer of electrons from semiconductor to metal) (Figure 19a), metal SPR with transfer of charge from metal to semiconductor (Figure 19b) and metal SPR—local electric field (accompanied by recombination of electrons from metal and holes of semiconductors) (Figure 19c). The SPR for Ag NPs is observed generally near the wave-length of 400 nm, while adsorption of Au NPs is observed at 550 nm [181], which makes gold more attractive for photocatalysis [182,183]. However, the position of the absorption band of nanoparticles depends on many factors, including the size and shape of particles, interaction with surroundings. Thus, significant shift of SPR of Ag NPs from 400 nm to 480–500 nm is observed for Ag/CeO$_2$ catalysts [184] that may be attributed to strong electronic metal–support interaction between Ag and CeO$_2$. This provides an enhanced photocatalytic activity of Ag/CeO$_2$ composites in the degradation of methylene blue under the simulated sunlight [50] or visible light [185]. According to [50], Ag acts as an acceptor of photoelectrons, and then the electron rapidly reacts with O$_2$ yielding O$_2^-$ that reduces the probability of recombination of electron–hole pairs. The correlation between the rate of degradation and amount of Ag NPs (active sites) was found. High stability and high recyclability of the Ag/CeO$_2$ heterostructure catalysts was shown.

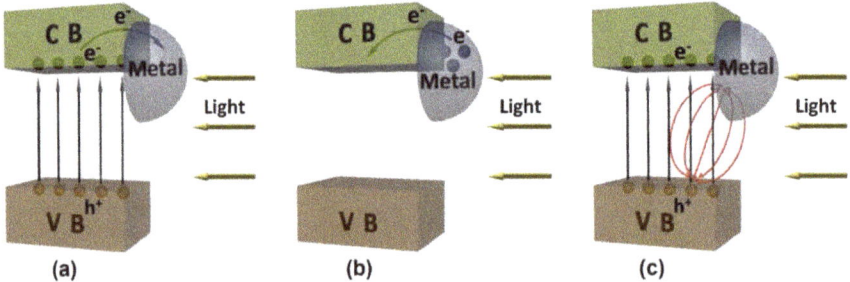

Figure 19. Schematic diagram showing: (**a**) transfer of electrons to form a Schottky barrier (**b**) transfer of electrons excited by surface plasmon resonance, (**c**) excitation of electrons in the photocatalyst from the local electric field. Reproduced from Ref. [170] with the permission from the Elsevier.

In Ref. [186], a photocatalytic degradation of Congo Red under UV light and visible light over three-dimensionally ordered macroporous (3DOM) Ag/CeO$_2$–ZrO$_2$ material was studied. It was shown that the SPR effect of Ag particles provides the adsorption of visible light and promotes separation of electrons and holes, reducing their recombination and improving the photocatalytic activity. The superior photocatalytic activity of Ag/CeO$_2$/ZnO nanostructure was shown in degradation of azo dyes (methylene orange and methylene blue) and phenol solution under visible light irradiation was demonstrated in Ref. [187]. It was found that formation of oxygen vacancies led to a narrow band gap (2.66 eV), which helps to produce sufficient electrons and holes under visible light in the ternary Ag/CeO$_2$/ZnO nanostructure. The defect structure of composite inhibited the electron–hole recombination and provided synergistic effect of narrow band gap. The SPR of Ag NPs and defects (Ce^{3+} and oxygen vacancies) in CeO$_2$ and ZnO resulted in superior photocatalytic activity. In Ref. [188], the correlation between Ce^{3+} loading, amount of oxygen vacancies and activity of Ag/CeO$_2$ and Au/CeO$_2$ catalysts in photodegradation of rhodamine blue dye in an aqueous medium under UV–vis irradiations were found. The conditions of synthesis (pH of precipitation) and Ag/Au loading provided different Ce^{3+} loading, distortion of CeO$_2$ lattice and concentration of vacancies. All these parameters affected on light absorbance, separation of photogenerated charges and photocatalytic properties.

Thus, silver and its compounds supported on ceria have high importance in photocatalytic degradation of organic pollutants. Semiconductor properties of silver compounds and SPR of Ag NPs provide both absorbance of visible light, separation of electrons and holes and result in increased photocatalytic activity. Several common aspects were found between classical oxidation catalysis and photocatalysis over Ag/CeO$_2$ composites. The interaction of silver with ceria (including electronic metal–support interaction) influences on the catalytic activity of Ag/CeO$_2$ due to cooperation of active sites of Ag and ceria. The presence of Ag–CeO$_2$ contact also leads to a growth of the amount of oxygen vacancies in the structure of CeO$_2$ that also promotes an enhanced catalytic/photocatalytic activity. Generally, Ag/CeO$_2$ composites are new for photocatalysis and poorly described. The study of Ag/CeO$_2$ systems in photocatalysis has high importance for fundamental research and real application of catalysts in the purification of aqueous wastes from dyes and other organic pollutants.

4.2. Electrocatalysis

Silver was also shown to be a promising material for electrocatalytic applications [189–191]. Recently, ceria has attracted a growing interest as a component of materials for electrocatalytic applications [192,193]. The main reasons for this are its high oxygen storage and transfer abilities. Application of proper amounts of noble metal improves the conductive properties of CeO_2-based materials, thus making them promising composites for electrocatalytic applications in fuel cells, metal-air batteries, and other alternative energy transfer devices [194].

A combination of silver and ceria in Ag/CeO_2 composites was used in several publications [51,52,195,196]. In Ref. [196] the Ag/CeO_2 composites comprising 30–50 nm silver nanoparticles uniformly anchored on the surface of nanosheet-constructing porous CeO_2 microspheres were used as oxygen reduction reaction catalysts. CeO_2 is known to show high oxygen storage capability and oxygen transfer ability, and silver was added to improve the conductivity of the latter. As a result, an enhanced activity was observed, and aluminum–air batteries based on Ag/CeO_2 composites exhibited an output power density of 345 mW/cm^2 and low degradation rate of 2.6% per 100 h, respectively.

In Ref. [51] a method was developed to prepare nanoporous $Ag–CeO_2$ ribbons with a homogeneous pore/grain structure by dealloying melt-spun Al–Ag–Ce alloy in a 5 wt. % NaOH aqueous solution. The resulting structure comprised uniform CeO_2 particles dispersed on the fine Ag grains, with the amount of oxygen vacancies growing as the calcination temperature increases. An enhanced $Ag–CeO_2$ interfacial interaction was assumed to cause high performance of the composites in electrocatalytic oxidation of sodium borohydride. In Ref. [195] Au was shown to impair the promoting effect on these composites and decrease the reaction resistance. The activity improvement was assumed to be caused by strengthening of interfacial interaction between the Ag–Au solid solution and CeO_2 particles due to Au effect, while the thermal stability and electron transport properties also improved. An increase of the Au content in the precursor alloy results in the reduction of catalytic activity and thermal stability.

In Ref. [52] 3D Ag/CeO_2 nanorods with high electrocatalytic activity for $NaBH_4$ electrooxidation were discussed. The ongoing calcination in air resulted in the dispersion of small Ag nanoparticles on the CeO_2 surface, and well-defined $Ag–CeO_2$ interfaces were created, where nanorods were connected by large conductive Ag nanoparticles. The resulting mass specific current of the composite 2.5 times exceeded the one for pure Ag in borohydride oxidation reaction. High concentration of surface oxygen species was assumed to determine the exhibited enhanced catalytic activity along with a 3D architecture of nanorod and strong metal–support interaction.

Thus, a variation of the chemical composition of Ag/CeO_2 by using various promoters and modifiers allows tuning the electrocatalytic activity of the composite.

5. Conclusions and Outlook

In the present review we have summarized the recent advances and trends on the role of metal–support interaction in Ag/CeO_2 composites in their catalytic performance in total oxidation of CO, soot, and VOCs. Promising photo- and electrocatalytic applications of Ag/CeO_2 composites have also been discussed. The key function of the silver–ceria interaction is connected with the following major aspects:

1. the catalytic performance of Ag/CeO_2 composites strongly depends on the preparation method that determines the morphology of both Ag and ceria nanoparticles, interfacial configuration and strength of metal–support interaction;
2. active surface sites are formed at the $Ag–CeO_2$ interface, with the interfacial O atoms exhibiting different reactivity as compared to other surface O atoms, while oxygen species over Ag particles are still of importance and participate in catalysis;

3. positively charged Ag clusters facilitate the formation of surface oxygen vacancies over ceria support, while metal Ag nanoparticles promote the reduction of CeO_2 nanocrystals and enhance their catalytic activity;
4. an enhanced activity of Ag/CeO_2 materials is caused by the highest surface oxygen vacancy concentration, high low-temperature reducibility as well as existence of lattice oxygen species and lattice defects formed with the participation of both silver and ceria;
5. the role of impurities (such as alkali ions, carbon-containing species, etc., appeared on the surface and/or bulk of ceria during the preparation procedure and participating in transferring of electron density to O surface species) should be considered;
6. redox properties are caused by coexistence and interplay between Ag^+/Ag^0 and Ce^{3+}/Ce^{4+} pairs;
7. high photocatalytic activity of Ag/CeO_2 composites is caused by the ability of Ag nanoparticles to prolong the lifetime of photogenerated electron–hole pairs due to the effect of localized SPR and reduction of the recombination of free charges;
8. enhanced electrocatalytic activity and good electrochemical stability of Ag/CeO_2 composites are connected with strong interfacial interactions between Ag and CeO_2 moieties that are caused by their specific morphology and architecture, which hinder the particulate agglomeration during the long-term electrocatalytic reaction.

Thus, the configuration of the silver–ceria interface provides the enhanced catalyst performance caused by synergistic effects of silver and cerium oxide. A proper selection of preparation method allows achieving the desired features of the composites and fine-tuning the strength of electronic metal–support interactions that can be additionally improved by application of ordered supports (e.g., SBA, MCM, MOFs, etc.) and promoters. This will allow rational designing of a new generation of highly effective Ag/CeO_2 composites for environmental, energy, photo- and electrocatalytic applications.

Acknowledgments: This research was supported by "The Tomsk State University competitiveness improvement programme".

Conflicts of Interest: The authors declare no conflict of interest.

References

1. Web Site of World Health Organization. Available online: http://www.who.int/mediacentre/factsheets/fs313/en/ (accessed on 2 May 2018).
2. California Air Resources Board. *Definitions of VOC and ROG*; California Air Resources Board: Sacramento, CA, USA, 2004.
3. Kamal, M.S.; Razzak, S.A.; Hossain, M.M. Catalytic oxidation of volatile organic compounds (VOCs)-A review. *Atmos. Environ.* **2016**, *140*, 117–134. [CrossRef]
4. Oh, S.-H.; Hoflund, G.B. Chemical state study of palladium powder and ceria-supported palladium during low-temperature CO oxidation. *J. Phys. Chem. A* **2006**, *110*, 7609–7613. [CrossRef] [PubMed]
5. Slavinskaya, E.M.; Gulyaev, R.V.; Zadesenets, A.V.; Stonkus, O.A.; Zaikovskii, V.I.; Shubin, Y.V.; Korenev, S.V.; Boronin, A.I. Low-temperature CO oxidation by Pd/CeO_2 catalysts synthesized using the coprecipitation method. *Appl. Catal. B* **2015**, *166–167*, 91–103. [CrossRef]
6. Hinokuma, S.; Fujii, H.; Okamoto, M.; Ikeue, K.; Machida, M. Metallic Pd nanoparticles formed by Pd–O–Ce interaction: a reason for sintering-induced activation for CO oxidation. *Chem. Mater.* **2010**, *22*, 6183–6190. [CrossRef]
7. Jin, M.; Park, J.-N.; Shon, J.K.; Kim, J.H.; Li, Z.; Park, Y.-K.; Kim, J.M. Low temperature CO oxidation over Pd catalysts supported on highly ordered mesoporous metal oxides. *Catal. Today* **2012**, *185*, 183–190. [CrossRef]
8. Wu, J.; Zeng, L.; Cheng, D.; Chen, F.; Zhan, X.; Gong, J. Synthesis of Pd nanoparticles supported on CeO_2 nanotubes for CO oxidation at low temperatures. *Chin. J. Catal.* **2016**, *37*, 83–90. [CrossRef]
9. Hu, Z.; Liu, X.F.; Meng, D.M.; Guo, Y.; Guo, Y.L.; Lu, G.Z. Effect of Ceria Crystal Plane on the Physicochemical and Catalytic Properties of Pd/Ceria for CO and Propane Oxidation. *ACS Catal.* **2016**, *6*, 2265–2279. [CrossRef]

10. Si, G.; Yu, J.; Xiao, X.; Guo, X.; Huang, H.; Mao, D.; Lu, G. Boundary role of Nano-Pd catalyst supported on ceria and the approach of promoting the boundary effect. *Mol. Catal.* **2018**, *444*, 1–9. [CrossRef]
11. Adijanto, L.; Sampath, A.; Yu, A.S.; Cargnello, M.; Fornasiero, P.; Gorte, R.J.; Vohs, J.M. Synthesis and Stability of Pd@CeO_2 Core–Shell Catalyst Films in Solid Oxide Fuel Cell Anodes. *ACS Catal.* **2013**, *3*, 1801–1809. [CrossRef]
12. Carlsson, P.A.; Skoglundh, M. Low-temperature oxidation of carbon monoxide and methane over alumina and ceria supported platinum catalysts. *Appl. Catal. B* **2011**, *101*, 669–675. [CrossRef]
13. Morfin, F.; Nguyen, T.S.; Rousset, J.L.; Piccolo, L. Synergy between hydrogen and ceria in Pt-catalyzed CO oxidation: An investigation on Pt–CeO_2 catalysts synthesized by solution combustion. *Appl. Catal. B* **2016**, *197*, 2–13. [CrossRef]
14. Lee, J.; Ryou, Y.; Kim, J.; Chan, X.; Kim, T.J.; Kim, D.H. Influence of the Defect Concentration of Ceria on the Pt Dispersion and the CO Oxidation Activity of Pt/CeO_2. *J. Phys. Chem. C* **2018**, *122*, 4972–4983. [CrossRef]
15. Peng, R.; Li, S.; Sun, X.; Ren, Q.; Chen, L.; Fu, M.; Wu, J.; Ye, D. Size effect of Pt nanoparticles on the catalytic oxidation of toluene over Pt/CeO_2 catalysts. *Appl. Catal. B* **2018**, *220*, 462–470. [CrossRef]
16. Bera, P.; Gayen, A.; Hegde, M.S.; Lalla, N.P.; Spadaro, L.; Frusteri, F.; Arena, F. Promoting Effect of CeO_2 in Combustion Synthesized Pt/CeO_2 Catalyst for CO Oxidation. *J. Phys. Chem. B* **2003**, *107*, 6122–6130. [CrossRef]
17. Zhang, R.; Lu, K.; Zong, L.; Tong, S.; Wang, X.; Feng, G. Gold supported on ceria nanotubes for CO oxidation. *Appl. Surf. Sci.* **2017**, *416*, 183–190. [CrossRef]
18. Zhang, R.; Lu, K.; Zong, L.; Tong, S.; Wang, X.; Zhou, J.; Lu, Z.-H.; Feng, G. CO Oxidation Activity at Room Temperature over Au/CeO_2 Catalysts: Disclosure of Induction Period and Humidity Effect. *Mol. Catal.* **2017**, *442*, 173–180. [CrossRef]
19. Centeno, M.A.; Reina, T.R.; Ivanova, S.; Laguna, O.H.; Odriozola, J.A. Au/CeO_2 Catalysts: Structure and CO Oxidation Activity. *Catalysts* **2016**, *6*, 158. [CrossRef]
20. El-Moemen, A.A.; Abdel-Mageed, A.M.; Bansmann, J.; Parlinska-Wojtan, M.; Behm, R.J.; Kučerová, G. Deactivation of Au/CeO_2 catalysts during CO oxidation: Influence of pretreatment and reaction conditions. *J. Catal.* **2016**, *341*, 160–179. [CrossRef]
21. Sudarsanam, P.; Mallesham, B.; Reddy, P.S.; Großmann, D.; Grünert, W.; Reddy, B.M. Nano-Au/CeO_2 catalysts for CO oxidation: Influence of dopants (Fe, La and Zr) on the physicochemical properties and catalytic activity. *Appl. Catal. B* **2014**, *144*, 900–908. [CrossRef]
22. Zhang, S.; Li, X.-S.; Chen, B.; Zhu, X.; Shi, C.; Zhu, A.-M. CO Oxidation Activity at Room Temperature over Au/CeO_2 Catalysts: Disclosure of Induction Period and Humidity Effect. *ACS Catal.* **2014**, *4*, 3481–3489. [CrossRef]
23. Li, H.-F.; Zhang, N.; Chen, P.; Luo, M.-F.; Lu, J.-Q. High surface area Au/CeO2 catalysts for low temperature formaldehyde oxidation. *Appl. Catal. B* **2011**, *110*, 279–285. [CrossRef]
24. Satsuma, A.; Yanagihara, M.; Ohyama, J.; Shimizu, K. Oxidation of CO over Ru/ceria prepared by self-dispersion of Ru metal powder into nano-sized particle. *Catal. Today* **2013**, *201*, 62–67. [CrossRef]
25. Vargas, E.; Simakov, A.; Rangel, R.; Castillon, F. CO oxidation over Ce–Ru–O catalysts. In Proceedings of the 20th North American Catalysis Meeting, Houston, TX, USA, 17–22 June 2007.
26. Asadullah, M.; Fujimoto, K.; Tomishige, K. Catalytic Performance of Rh/CeO_2 in the Gasification of Cellulose to Synthesis Gas at Low Temperature. *Ind. Eng. Chem. Res.* **2001**, *40*, 5894–5900. [CrossRef]
27. Li, K.; Wang, X.; Zhou, Z.; Wu, X.; Weng, D. Oxygen Storage Capacity of Pt-, Pd-, Rh/CeO_2-Based Oxide Catalyst. *J. Rare Earths* **2007**, *25*, 6–10.
28. Kurnatowska, M.; Kepinski, L. Structure and thermal stability of nanocrystalline $Ce_{1-x}Rh_xO_{2-y}$ in reducing and oxidizing atmosphere. *Mater. Res. Bull.* **2013**, *48*, 852–862. [CrossRef]
29. Zhao, Y.; Teng, B.-T.; Wen, X.-D.; Zhao, Y.; Zhao, L.-H.; Luo, M.-F. A theoretical evaluation and comparison of $M_xCe_{1-x}O_{2-\delta}$ (M = Au, Pd, Pt, and Rh) catalysts. *Catal. Commun.* **2012**, *27*, 63–68. [CrossRef]
30. Li, Y.; Cai, Y.; Xing, X.; Chen, N.; Deng, D.; Wang, Y. Catalytic activity for CO oxidation of Cu–CeO_2 composite nanoparticles synthesized by a hydrothermal method. *Anal. Methods* **2015**, *7*, 3238–3245. [CrossRef]
31. Sundar, R.S.; Deevi, S. CO oxidation activity of Cu–CeO_2 nano-composite catalysts prepared by laser vaporization and controlled condensation. *J. Nanopart. Res.* **2006**, *8*, 497–509. [CrossRef]
32. Xu, X.; Li, J.; Hao, Z. CeO_2-Co_3O_4 Catalysts for CO Oxidation. *J. Rare Earths* **2006**, *24*, 172–176. [CrossRef]

33. Quiroz, J.; Giraudon, J.-M.; Gervasini, A.; Dujardin, C.; Lancelot, C.; Trentesaux, M.; Lamonier, J.-F. Total Oxidation of Formaldehyde over MnOx-CeO$_2$ Catalysts: The Effect of Acid Treatment. *ACS Catal.* **2015**, *5*, 2260–2269. [CrossRef]
34. Huang, H.; Xu, Y.; Feng, Q.; Leung, D.Y. Low temperature catalytic oxidation of volatile organic compounds: A review. *Catal. Sci. Technol.* **2015**, *5*, 2649–2669. [CrossRef]
35. Vodyankina, O.V.; Blokhina, A.S.; Kurzina, I.A.; Sobolev, V.I.; Koltunov, K.Y.; Chukhlomina, L.N.; Dvilis, E.S. Selective oxidation of alcohols over Ag-containing Si$_3$N$_4$ catalysts. *Catal. Today* **2013**, *203*, 127–132. [CrossRef]
36. Mamontov, G.V.; Grabchenko, M.V.; Sobolev, V.I.; Zaikovskii, V.I.; Vodyankina, O.V. Ethanol dehydrogenation over Ag-CeO$_2$/SiO$_2$ catalyst: Role of Ag-CeO$_2$ interface. *Appl. Catal. A* **2016**, *528*, 161–167. [CrossRef]
37. Dutov, V.V.; Mamontov, G.V.; Sobolev, V.I.; Vodyankina, O.V. Silica-supported silver-containing OMS-2 catalysts for ethanol oxidative dehydrogenation. *Catal. Today* **2016**, *278*, 164–173. [CrossRef]
38. Mamontov, G.V.; Gorbunova, A.S.; Vyshegorodtseva, E.V.; Zaikovskii, V.I.; Vodyankina, O.V. Selective oxidation of CO in the presence of propylene over Ag/MCM-41 catalyst. *Catal. Today* **2018**. [CrossRef]
39. Pan, C.-J.; Tsai, M.-C.; Su, W.-N.; Rick, J.; Akalework, N.G.; Agegnehu, A.K.; Cheng, S.-Y.; Hwang, B.-J. Tuning/exploiting Strong Metal-Support Interaction (SMSI) in Heterogeneous Catalysis. *J. Taiwan Inst. Chem. E* **2017**, *74*, 154–186. [CrossRef]
40. Ma, L.; Wang, D.; Li, J.; Bai, B.; Fu, L.; Li, Y. Ag/CeO$_2$ nanospheres: Efficient catalysts for formaldehyde oxidation. *Appl. Catal. B* **2014**, *148*, 36–43. [CrossRef]
41. Skaf, M.; Aouad, S.; Hany, S.; Cousin, R.; Abi-Aadand, E.; Aboukais, A. Physicochemical characterization and catalytic performance of 10% Ag/CeO$_2$ catalysts prepared by impregnation and deposition-precipitation. *J. Catal.* **2014**, *320*, 137–146. [CrossRef]
42. Qu, Z.; Yu, F.; Zhang, X.; Wang, Y.; Gao, J. Support effects on the structure and catalytic activity of mesoporous Ag/CeO$_2$ catalysts for CO oxidation. *Chem. Eng. J.* **2013**, *229*, 522–532. [CrossRef]
43. Chang, S.; Li, M.; Hua, Q.; Zhang, L.; Ma, Y.; Ye, B.; Huang, W. Shape-dependent interplay between oxygen vacancies and Ag–CeO2 interaction in Ag/CeO$_2$ catalysts and their influence on the catalytic activity. *J. Catal.* **2012**, *293*, 195–204. [CrossRef]
44. Kayama, T.; Yamazaki, K.; Shinjoh, H. Nanostructured Ceria−Silver Synthesized in a One-Pot Redox Reaction Catalyzes Carbon Oxidation. *J. Am. Chem. Soc.* **2010**, *132*, 13154–13155.
45. Shimizu, K.; Kawachi, H.; Satsuma, A. Study of active sites and mechanism for soot oxidation by silver-loaded ceria catalyst. *Appl. Catal. B Environ.* **2010**, *96*, 169–175. [CrossRef]
46. Aneggi, E.; Llorca, J.; de Leitenburg, C.; Dolcetti, G.; Trovarelli, A. Soot Combustion Over Silver-Supported Catalysts. *Appl. Catal. B Environ.* **2009**, *91*, 489–498. [CrossRef]
47. Lee, C.; Park, J.; Shul, Y.-G.; Einaga, H.; Teraoka, Y. Ag supported on electrospun macro-structure CeO$_2$ fibrous mats for diesel soot oxidation. *Appl. Catal. B Environ.* **2015**, *174*, 185–192. [CrossRef]
48. Machida, M.; Murata, Y.; Kishikawa, K.; Zhang, D.; Ikeue, K. On the Reasons for High Activity of CeO$_2$ Catalyst for Soot Oxidation. *Chem. Mater.* **2008**, *20*, 4489–4494. [CrossRef]
49. Yamazaki, K.; Kayama, T.; Dong, F.; Shinjoh, H. A mechanistic study on soot oxidation over CeO$_2$-Ag catalyst with 'rice-ball' morphology. *J. Catal.* **2011**, *282*, 289–298. [CrossRef]
50. Leng, Q.; Yang, D.; Yang, Q.; Hu, C.; Kang, Y.; Wang, M.; Hashim, M. Building novel Ag/CeO$_2$ heterostructure for enhancing photocatalytic activity. *Mater. Res. Bull.* **2015**, *65*, 266–272. [CrossRef]
51. Li, G.; Lu, F.; Wei, X.; Song, X.; Sun, Z.; Yang, Z.; Yang, S. Nanoporous Ag–CeO$_2$ ribbons prepared by chemical dealloying and their electrocatalytic properties. *J. Mater. Chem.* **2013**, *1*, 4974–4981. [CrossRef]
52. Zhang, X.; Li, G.; Song, X.; Yang, S.; Sun, Z. Three-dimensional architecture of Ag/CeO$_2$ nanorod composites prepared by dealloying and their electrocatalytic performance. *RSC Adv.* **2017**, *7*, 32442–32451. [CrossRef]
53. Imamura, S.; Yamada, H.; Utani, K. Combustion activity of Ag/CeO$_2$ composite catalyst. *Appl. Catal. A Gen.* **2000**, *192*, 221–226. [CrossRef]
54. Scirè, S.; Riccobene, P.M.; Crisafulli, C. Ceria supported group IB metal catalysts for the combustion of volatile organic compounds and the preferential oxidation of CO. *Appl. Catal. B Environ.* **2010**, *101*, 109–117. [CrossRef]
55. Fiorenza, R.; Crisafulli, C.; Condorelli, G.G.; Lupo, F.; Scirè, S. Au–Ag/CeO$_2$ and Au–Cu/CeO$_2$ Catalysts for Volatile Organic Compounds Oxidation and CO Preferential Oxidation. *Catal. Lett.* **2015**, *145*, 1691–1702. [CrossRef]

56. Wang, L.; He, H.; Yu, Y.; Sun, L.; Liu, S.; Zhang, C.; He, L. Morphology-dependent bactericidal activities of Ag/CeO$_2$ catalysts against Escherichia coli. *J. Inorg. Biochem.* **2014**, *135*, 45–53. [CrossRef] [PubMed]
57. Beuhler, R.J.; Rao, R.M.; Hrbek, J.; White, M.G. Study of the Partial Oxidation of Methanol to Formaldehyde on a Polycrystalline Ag Foil. *J. Phys. Chem. B* **2001**, *105*, 5950–5956. [CrossRef]
58. Mamontov, G.V.; Magaev, O.V.; Knyazev, A.S.; Vodyankina, O.V. Influence of phosphate addition on activity of Ag and Cu catalysts for partial oxidation of alcohols. *Catal. Today* **2013**, *203*, 122–126. [CrossRef]
59. Rodriguez, J.A.; Grinter, D.C.; Liu, Z.; Palomino, R.M.; Senanayake, S.D. Ceria-based model catalysts: Fundamental studies on the importance of the metal-ceria interface in CO oxidation, the water-gas shift, CO$_2$ hydrogenation, and methane and alcohol reforming. *Chem. Soc. Rev.* **2017**, *46*, 1824–1841. [CrossRef] [PubMed]
60. Zagaynov, I.V.; Naumkin, A.V.; Grigoriev, Y.V. Perspective intermediate temperature ceria based catalysts for CO oxidation. *Appl. Catal. B Environ.* **2018**, *236*, 171–175. [CrossRef]
61. Slavinskaya, E.M.; Stadnichenko, A.I.; Muravyov, V.V.; Kardash, T.Y.; Derevyannikova, E.A.; Zaikovskii, V.I.; Stonkus, O.A.; Lapin, I.N.; Svetlichnyi, V.A.; Boronin, A.I. Transformation of a Pt–CeO$_2$ Mechanical Mixture of Pulsed-Laser-Ablated Nanoparticles to a Highly Active Catalyst for Carbon Monoxide Oxidation. *ChemCatChem* **2018**, *10*, 2232–2247. [CrossRef]
62. Park, Y.; Kim, S.K.; Pradhan, D.; Sohn, Y. Thermal H$_2$-treatment effects on CO/CO$_2$ conversion over Pd-doped CeO$_2$ comparison with Au and Ag-doped CeO$_2$. *Reac. Kinet. Mech. Cat.* **2014**, *113*, 85–100. [CrossRef]
63. Park, Y.; Na, Y.; Pradhan, Y.; Sohn, Y. Liquid-Phase Ethanol Oxidation and Gas-Phase CO Oxidation Reactions over M Doped (M = Ag, Au, Pd, and Ni) and MM' Codoped CeO$_2$ Nanoparticles. *J. Catal.* **2016**, 2176576. [CrossRef]
64. Dutov, V.V.; Mamontov, G.V.; Zaikovskii, V.I.; Liotta, L.F.; Vodyankina, O.V. Low-temperature CO oxidation over Ag/SiO$_2$ catalysts: Effect of OH/Ag ratio. *Appl. Catal. B* **2018**, *221*, 598–609. [CrossRef]
65. Dutov, V.V.; Mamontov, G.V.; Zaikovskii, V.I.; Vodyankina, O.V. The effect of support pretreatment on activity of Ag/SiO$_2$ catalysts in low-temperature CO oxidation. *Catal. Today* **2016**, *278*, 150–156. [CrossRef]
66. Afanasev, D.S.; Yakovina, O.A.; Kuznetsova, N.I.; Lisitsyn, A.S. High activity in CO oxidation of Ag nanoparticles supported on fumed silica. *Catal. Commun.* **2012**, *22*, 43–47. [CrossRef]
67. Liu, H.; Ma, D.; Blackley, R.A.; Zhou, W.; Bao, X. Highly active mesostructured silica hosted silver catalysts for CO oxidation using the one-pot synthesis approach. *Chem. Commun.* **2008**, 2677–2678. [CrossRef] [PubMed]
68. Mamontov, G.V.; Dutov, V.V.; Sobolev, V.I.; Vodyankina, O.V. Effect of transition metal oxide additives on the activity of an Ag/SiO2 catalyst in carbon monoxide oxidation. *Kinet. Catal.* **2013**, *54*, 487–491. [CrossRef]
69. Zhang, D.; Du, X.; Shi, L.; Gao, R. Shape-controlled synthesis and catalytic application of ceria nanomaterials. *Dalton Trans.* **2012**, *41*, 14455–14475. [CrossRef] [PubMed]
70. Mai, H.-X.; Sun, L.-D.; Zhang, Y.-W.; Si, R.; Feng, W.; Zhang, H.-P.; Liu, H.-C.; Yan, C.-H. Shape-Selective Synthesis and Oxygen Storage Behavior of Ceria Nanopolyhedra, Nanorods, and Nanocubes. *J. Phys. Chem. B.* **2005**, *109*, 24380–24385. [CrossRef] [PubMed]
71. Wu, Z.; Li, M.; Overbury, S.H. On the structure dependence of CO oxidation over CeO$_2$ nanocrystals with well-defined surface planes. *J. Catal.* **2012**, *285*, 61–73. [CrossRef]
72. Kang, Y.; Sun, M.; Li, A. Studies of the Catalytic Oxidation of CO Over Ag/CeO$_2$ Catalyst. *Catal. Lett.* **2012**, *142*, 1498–1504. [CrossRef]
73. Li, G.; Zhang, X.; Feng, W.; Fang, X.; Liu, J. Nanoporous CeO$_2$–Ag catalysts prepared by etching the CeO$_2$/CuO/Ag$_2$O mixed oxides for CO oxidation. *Corros. Sci.* **2018**, *134*, 140–148. [CrossRef]
74. Liang, X.; Xiao, J.; Chen, B.; Li, Y. Catalytically Stable and Active CeO$_2$ Mesoporous Spheres. *Inorg. Chem.* **2010**, *49*, 8188–8190. [CrossRef] [PubMed]
75. Li, G.; Tang, Z. Noble metal nanoparticle@metal oxide core/yolk–shell nanostructures as catalysts: recent progress and perspective. *Nanoscale* **2014**, *6*, 3995–4011. [CrossRef] [PubMed]
76. Wang, Y.; Arandiyan, H.; Scott, J.; Bagheri, A.; Dai, H.; Amal, R. Recent Advances in Ordered Meso/macroporous Metal Oxides for Heterogeneous Catalysis: A Review. *J. Mater. Chem. A* **2017**, *5*, 8825–8846. [CrossRef]

77. Zhang, J.; Li, L.; Huang, X.; Li, G. Fabrication of Ag–CeO$_2$ core–shell nanospheres with enhanced catalytic performance due to strengthening of the interfacial interactions. *J. Mater. Chem.* **2012**, *22*, 10480–10487. [CrossRef]
78. Badri, A.; Binet, C.; Lavalley, J.-C. An FTIR study of surface ceria hydroxy groups during a redox process with H$_2$. *J. Chem. Soc. Faraday Trans.* **1996**, *92*, 4669–4673. [CrossRef]
79. Grabchenko, M.V.; Mamontov, G.V.; Zaikovskii, V.I.; La Parola., V.; Liotta, L.F.; Vodyankina, O.V. Design of Ag-CeO$_2$/SiO$_2$ catalyst for oxidative dehydrogenation of ethanol: Control of Ag–CeO$_2$ interfacial interaction. *Catal. Today* **2018**. [CrossRef]
80. Grabchenko, M.V.; Mamontov, G.V.; Zaikovskii, V.I.; Vodyankina, O.V. Effect of the metal−support interaction in Ag/CeO$_2$ catalysts on their activity in ethanol oxidation. *Kinet. Catal.* **2017**, *58*, 642–648. [CrossRef]
81. Mitsudome, T.; Mikami, Y.; Matoba, M.; Mizugaki, T.; Jitsukawa, K.; Kaneda, K. Design of a silver-cerium dioxide core-shell nanocomposite catalyst for chemoselective reduction reactions. *Angew. Chem. Int. Ed.* **2012**, *51*, 136–139. [CrossRef] [PubMed]
82. Zhou, Q.; Ma, S.; Zhan, S. Superior photocatalytic disinfection effect of Ag-3D ordered mesoporous CeO$_2$ under visible light. *Appl. Catal. B* **2018**, *224*, 27–37. [CrossRef]
83. Stanmore, B.R.; Brilhac, J.F.; Gilot, P. The oxidation of soot: a review of experiments, mechanisms and models. *Carbon* **2001**, *39*, 2247–2268. [CrossRef]
84. Neyertz, C.A.; Banus, E.D.; Mir'o, E.E.; Querini, C.A. Potassium-promoted Ce$_{0.65}$Zr$_{0.35}$O$_2$ monolithic catalysts for diesel soot combustion. *Chem. Eng. J.* **2014**, *248*, 394–405. [CrossRef]
85. Fino, D.; Bensaid, S.; Piumetti, M.; Russo, N. A review on the catalyticcombustion of soot in diesel particulate filters for automotive applications:from powder catalysts to structured reactors. *Appl. Catal. A* **2016**, *509*, 75–96. [CrossRef]
86. Neeft, P.A.; Makkee, M.; Moulijn, J.A. Catalysts for the oxidation of soot from diesel exhaust gases. I. An exploratory study. *Appl. Catal. B* **1996**, *8*, 57–78. [CrossRef]
87. Kobayashi, Y.; Hikosaka, R. Analyzing Loose Contact Oxidation of Diesel Engine Soot and Ag/CeO$_2$ Catalyst Using Nonlinear Regression Analysis. *Bull. Chem. React. Eng. Catal.* **2017**, *12*, 14–23. [CrossRef]
88. Piumetti, M.; Bensaid, S.; Russo, N.; Fino, D. Nanostructured ceria-based catalysts for soot combustion: Investigations on the surface sensitivity. *Appl. Catal. B* **2015**, *165*, 742–751. [CrossRef]
89. Hernández-Giménez, A.M.; Castelló, D.L.; Bueno-López, A. Diesel soot combustion catalysts: Review of active phases. *Chem. Pap.* **2014**, *68*, 1154–1168. [CrossRef]
90. Shen, Q.; Wu, M.; Wang, H.; He, C.; Hao, Z.; Wei, W.; Sun, Y. Facile synthesis of catalytically active CeO$_2$ for soot combustion. *Catal. Sci. Technol.* **2015**, *5*, 1941–1952. [CrossRef]
91. Gnanamani, M.K.; Jacobs, G.; Martinelli, M.; Shafer, W.D.; Hopps, S.D.; Thomas, G.A.; Davis, B.H. Dehydration of 1,5-Pentanediol over Na-Doped CeO$_2$ Catalysts. *ChemCatChem* **2018**, *10*, 1148–1154. [CrossRef]
92. Miceli, P.; Bensaid, S.; Russo, N.; Fino, D. CeO$_2$-based catalysts with engineered morphologies for soot oxidation to enhance soot-catalyst contact. *Nanoscale Res. Lett.* **2014**, *9*, 254–264. [CrossRef] [PubMed]
93. Guillén-Hurtado, N.; Bueno-López, A.; García-García, A. Catalytic performances of ceria and ceria-zirconia materials for the combustion of diesel soot under NO$_x$/O$_2$ and O$_2$. Importance of the cerium precursor salt. *Appl. Catal. A* **2012**, *437–438*, 166–172. [CrossRef]
94. Bensaid, S.; Russo, N.; Fino, D. CeO$_2$ catalysts with fibrous morphology for soot oxidation: The importance of the soot-catalyst contact conditions. *Catal. Today* **2013**, *216*, 57–63. [CrossRef]
95. Atribak, I.; Such-Basáñez, I.; Bueno-López, A.; García-García, A. Comparison of the catalytic activity of MO$_2$ (M = Ti, Zr, Ce) for soot oxidation under NO$_x$/O$_2$. *J. Catal.* **2007**, *250*, 75–84. [CrossRef]
96. Zhang, W.; Niu, X.; Chen, L.; Yuan, F.; Zhu, Y. Soot Combustion over Nanostructured Ceria with Different Morphologies. *Sci. Rep.* **2016**, *6*, 29062. [CrossRef] [PubMed]
97. Aneggi, E.; Wiater, D.; de Leitenburg, C.; Llorca, J.; Trovarelli, A. Shape-Dependent Activity of Ceria in Soot Combustion. *ACS Catal.* **2014**, *4*, 172–181. [CrossRef]
98. Kaspar, J.; Fornasiero, P.; Graziani, M. Use of CeO$_2$-based oxides in the three-way catalysis. *Catal. Today* **1999**, *50*, 285–298. [CrossRef]
99. Mukherjee, D.; Rao, B.G.; Reddy, B.M. CO and soot oxidation activity of doped ceria: Influence of dopants. *Appl. Catal. B* **2016**, *197*, 105–115. [CrossRef]

100. Mukherjeea, D.; Reddy, B.M. Noble metal-free CeO$_2$-based mixed oxides for CO and soot oxidation. *Catal. Today* **2018**, *309*, 227–235. [CrossRef]
101. Kong, D.; Wang, G.; Pan, Y.; Hu, S.; Hou, J.; Pan, H.; Campbell, C.T.; Zhu, J. Growth, Structure, and Stability of Ag on CeO$_2$(111): Synchrotron Radiation Photoemission Studies. *J. Phys. Chem. C* **2011**, *115*, 6715–6725. [CrossRef]
102. Liu, S.; Wu, X.; Weng, D.; . Ran, R. Ceria-based catalysts for soot oxidation: A review. *J. Rare Earth* **2015**, *33*, 567–590. [CrossRef]
103. Severin, N.; Kirstein, S.; Sokolov, I.M.; Rabe, J.P. Rapid trench channeling of graphenes with catalytic silver nanoparticles. *Nano Lett.* **2009**, *9*, 457–461. [CrossRef] [PubMed]
104. Yamazaki, K.; Sakakibara, Y.; Dong, F.; Shinjoh, H. The remote oxidation of soot separated by ash deposits viasilver–ceria composite catalysts. *Appl. Catal. A* **2014**, *476*, 113–120. [CrossRef]
105. Liu, S.; Wu, X.; Liu, W.; Chen, W.; Ran, R.; Li, M.; Weng, D. Soot oxidation over CeO$_2$ and Ag/CeO$_2$: Factors determining the catalyst activity and stability during reaction. *J. Catal.* **2016**, *337*, 188–198. [CrossRef]
106. Wu, S.; Yang, Y.; Lu, C.; Ma, Y.; Yuan, S.; Qian, G. Soot oxidation over CeO$_2$ or Ag/CeO$_2$: influences of bulk oxygen vacancies and surface oxygen vacancies on activity and stability of catalyst. *Eur. J. Inorg. Chem.* **2018**, *2018*, 2944–2951. [CrossRef]
107. Piumetti, M.; van der Linden, B.; Makkee, M.; Miceli, P.; Fino, D.; Russo, N.; Bensaid, S. Contact dynamics for a solid–solid reaction mediated by gas-phase oxygen: Study on the soot oxidation over ceria-based catalysts. *Appl. Catal. B* **2016**, *199*, 96–107. [CrossRef]
108. Shangguan, W.F.; Teraoka, Y.; Kagawa, S. Kinetics of soot–O$_2$, soot–NO and soot–O$_2$–NO reactions over spinel-type CuFe$_2$O$_4$ catalyst. *Appl. Catal. B* **1997**, *12*, 237–247. [CrossRef]
109. Aneggi, E.; Leitenburg, C.; Trovarelli, A. On the role of lattice/surface oxygen in ceria-zirconia catalysts for diesel soot combustion. *Catal. Today* **2012**, *181*, 108–115. [CrossRef]
110. Bueno-López, A. Diesel soot combustion ceria catalysts. *Appl. Catal. B* **2014**, *146*, 1–11. [CrossRef]
111. Bueno-López, A.; Krishna, K.; Makkee, M.; Moulijn, J.A. Enhanced soot oxidation by lattice oxygen via La^{3+}–doped CeO$_2$. *J. Catal.* **2005**, *230*, 237–248. [CrossRef]
112. Wang, H.; Liu, S.; Zhao, Z.; Zou, X.; Liu, M.; Liu, W.; Wu, X.; Weng, D. Activation and deactivation of Ag/CeO$_2$ during soot oxidation: influences of interfacial ceria reduction. *Catal. Sci. Technol.* **2017**, *7*, 2129–2139. [CrossRef]
113. Hosokawa, S.; Taniguchi, M.; Utani, K.; Kanai, H.; Imamura, S. Affinity order among noble metals and CeO$_2$. *Appl. Catal. A.* **2005**, *289*, 115–120. [CrossRef]
114. Nagai, Y.; Hirabayashi, T.; Dohmae, K.; Takagi, N.; Minami, T.; Shinjoh, H.; Matsumoto, S. Sintering inhibition mechanism of platinum supported on ceria-based oxide and Ptoxide–support interaction. *J. Catal.* **2006**, *242*, 103–109. [CrossRef]
115. Nagai, Y.; Dohmae, K.; Ikeda, Y.; Takagi, N.; Hara, N.; Tanabe, T.; Guilera, G.; Pascarelli, S.; Newton, M.A.; Takahashi, N.; et al. In situ observation of platinum sintering on ceria-based oxide for autoexhaust catalysts using Turbo-XAS. *Catal. Today* **2011**, *175*, 133–140. [CrossRef]
116. Malhautier, L.; Quijano, G.; Avezac, M.; Rocher, J.; Fanlo, J.L. Kinetic characterization of toluene biodegradation by Rhodococcus erythropolis: towards a rationale for microflora enhancement in bioreactors devoted to air treatment. *Chem. Eng. J.* **2014**, *247*, 199–204. [CrossRef]
117. Li, L.; Liu, S.; Liu, J. Surface modification of coconut shell based activated carbon for the improvement of hydrophobic VOC removal. *J. Hazard. Mater.* **2011**, *192*, 683–690. [CrossRef] [PubMed]
118. Thévenet, F.; Sivachandiran, L.; Guaitella, O.; Barakat, C.; Rousseau, A. Plasma–catalyst coupling for volatile organic compound removal and indoor air treatment: a review. *J. Phys. D Appl. Phys.* **2014**, *47*, 224011. [CrossRef]
119. Destaillats, H.; Sleiman, M.; Sullivan, D.P.; Jacquiod, C.; Sablayrolles, J.; Molins, L. Key parameters influencing the performance of photocatalytic oxidation (PCO) air purification under realistic indoor conditions. *Appl. Catal. B* **2012**, *128*, 159–170. [CrossRef]
120. Yuan, M.H.; Chang, C.Y.; Shie, J.L.; Chang, C.C.; Chen, J.H.; Tsai, W.T. Destruction of naphthalene via ozone-catalytic oxidation process over Pt/Al$_2$O$_3$ catalyst. *J. Hazard. Mater.* **2010**, *175*, 809–815. [CrossRef] [PubMed]
121. Liotta, L.F. Catalytic oxidation of volatile organic compounds on supported noble metals. *Appl. Catal. B* **2010**, *100*, 403–412. [CrossRef]

122. Everaert, K.; Baeyens, J. Catalytic combustion of volatile organic compounds. *J. Hazard. Mater.* **2004**, *109*, 113–139. [CrossRef] [PubMed]
123. Armor, J. N. Environmental catalysis. *Appl. Catal. B* **1992**, *1*, 221–256. [CrossRef]
124. Spivey, J.J. Complete catalytic oxidation of volatile organics. *Ind. Eng. Chem. Res.* **1987**, *26*, 2165–2180. [CrossRef]
125. Ordóñez, S.; Bello, L.; Sastre, H.; Rosal, R.; Dıez, F.V. Kinetics of the deep oxidation of benzene, toluene, n-hexane and their binary mixtures over a platinum on γ-alumina catalyst. *Appl. Catal. B* **2002**, *38*, 139–149. [CrossRef]
126. Huang, H.; Hu, P.; Huang, H.; Chen, J.; Ye, X.; Leung, D.Y. Highly dispersed and active supported Pt nanoparticles for gaseous formaldehyde oxidation: Influence of particle size. *Chem. Eng. J.* **2014**, *252*, 320–326. [CrossRef]
127. Huang, S.; Zhang, C.; He, H. Complete oxidation of o-xylene over Pd/Al_2O_3 catalyst at low temperature. *Catal. Today* **2008**, *139*, 15–23. [CrossRef]
128. Qi, J.; Chen, J.; Li, G.; Li, S.; Gao, Y.; Tang, Z. Facile synthesis of core–shell Au@ CeO_2 nanocomposites with remarkably enhanced catalytic activity for CO oxidation. *Energy Environ. Sci.* **2012**, *5*, 8937–8941. [CrossRef]
129. Hosseini, M.; Barakat, T.; Cousin, R.; Aboukaïs, A.; Su, B.L.; De Weireld, G.; Siffert, S. Catalytic performance of core–shell and alloy Pd–Au nanoparticles for total oxidation of VOC: the effect of metal deposition. *Appl. Catal. B* **2012**, *111*, 218–224. [CrossRef]
130. Xu, R.; Wang, X.; Wang, D.; Zhou, K.; Li, Y. Surface structure effects in nanocrystal MnO_2 and Ag/MnO_2 catalytic oxidation of CO. *J. Catal.* **2006**, *237*, 426–430. [CrossRef]
131. Bai, B.; Arandiyan, H.; Li, J. Comparison of the performance for oxidation of formaldehyde on nano-Co_3O_4, 2D-Co_3O_4, and 3D-Co_3O_4 catalysts. *Appl. Catal. B* **2013**, *142*, 677–683. [CrossRef]
132. Montini, T.; Melchionna, M.; Monai, M.; Fornasiero, P. Fundamentals and catalytic applications of CeO_2–based materials. *Chem. Rev.* **2016**, *116*, 5987–6041. [CrossRef] [PubMed]
133. Hanafiah, M.A.K.M.; Hussin, Z.M.; Ariff, N.F.M.; Ngah, W.S.W.; Ibrahim, S.C. Monosodium glutamate functionalized chitosan beads for adsorption of precious cerium ion. *Adv. Mater. Res.* **2014**, *970*, 198–203. [CrossRef]
134. Min, C.K. Nanostructured Pt/MnO2 Catalysts and Their Performance for Oxygen Reduction Reaction in Air Cathode Microbial Fuel Cell. Ph.D. Thesis, University Malaysia Pahang, Pekan, Pahang, Malaysia, June 2014.
135. Abdel-Mageed, A.M.; Kučerová, G.; El-Moemen, A.A.; Bansmann, J.; Widmann, D.; Behm, R.J. Geometric and electronic structure of Au on Au/CeO_2 catalysts during the CO oxidation: Deactivation by reaction induced particle growth. *J. Phys. Conf. Ser.* **2016**, *712*, 012044. [CrossRef]
136. Tan, H.; Wang, J.; Yu, S.; Zhou, K. Support morphology-dependent catalytic activity of Pd/CeO_2 for formaldehyde oxidation. *Environ. Sci. Technol.* **2015**, *49*, 8675–8682. [CrossRef] [PubMed]
137. Zhang, J.; Li, Y.; Zhang, Y.; Chen, M.; Wang, L.; Zhang, C.; He, H. Effect of support on the activity of Ag-based catalysts for formaldehyde oxidation. *Sci. Rep.* **2015**, *5*, 12950. [CrossRef] [PubMed]
138. Li, G.; Li, L. Highly efficient formaldehyde elimination over meso-structured M/CeO_2 (M=Pd, Pt, Au and Ag) catalyst under ambient conditions. *RSC Adv.* **2015**, *5*, 36428–36433. [CrossRef]
139. Yu, L.; Peng, R.; Chen, L.; Fu, M.; Wu, J.; Ye, D. Ag supported on CeO_2 with different morphologies for the catalytic oxidation of HCHO. *Chem. Eng. J.* **2018**, *334*, 2480–2487. [CrossRef]
140. Imamura, S.; Uchihori, D.; Utani, K.; Ito, T. Oxidative decomposition of formaldehyde on silver-cerium composite oxide catalyst. *Catal. Lett.* **1994**, *24*, 377–384. [CrossRef]
141. Benaissa, S.; Chérif-Aouali, L.; Siffert, S.; Aboukais, A.; Cousin, R.; Bengueddach, A. New Nanosilver/Ceria Catalyst for Atmospheric Pollution Treatment. *Nano* **2015**, *10*, 1550043. [CrossRef]
142. Aboukais, A.; Skaf, M.; Hany, S.; Cousin, R.; Aouad, S.; Labaki, M.; Abi-Aad, E. A comparative study of Cu, Ag and Au doped CeO_2 in the total oxidation of volatile organic compounds (VOCs). *Mater. Chem. Phys.* **2016**, *177*, 570–576. [CrossRef]
143. Liu, M.; Wu, X.; Liu, S.; Gao, Y.; Chen, Z.; Ma, Y.; Weng, D. Study of Ag/CeO_2 catalysts for naphthalene oxidation: Balancing the oxygen availability and oxygen regeneration capacity. *Appl. Catal. B* **2017**, *219*, 231–240. [CrossRef]
144. Yang, H.; Deng, J.; Liu, Y.; Xie, S.; Wu, Z.; Dai, H. Preparation and catalytic performance of Ag, Au, Pd or Pt nanoparticles supported on 3DOM CeO_2–Al_2O_3 for toluene oxidation. *J. Mol. Catal. A Chem.* **2016**, *414*, 9–18. [CrossRef]

145. Kharlamova, T.; Mamontov, G.; Salaev, M.; Zaikovskii, V.; Popova, G.; Sobolev, V.; Vodyankina, O. Silica-supported silver catalysts modified by cerium/manganese oxides for total oxidation of formaldehyde. *Appl. Catal. A* **2013**, *467*, 519–529. [CrossRef]
146. Mullins, D.R. The surface chemistry of cerium oxide. *Surf. Sci. Rep.* **2015**, *70*, 42–85. [CrossRef]
147. Nolan, N. Surface Effects in the Reactivity of Ceria: A First Principles Perspective. *Catal. Mater. Well-Defined Struct.* **2015**, 159–192.
148. Plata, J.J.; Márquez, A.M.; Fdez Sanz, J. Improving the density functional theory+U description of CeO_2 by including the contribution of the O 2p electrons. *J. Chem. Phys.* **2012**, *136*, 041101. [CrossRef] [PubMed]
149. Kozlov, S.M.; Viñes, F.; Nilius, N.; Shaikhutdinov, S.; Neyman, K.M. Absolute surface step energies: Accurate theoretical methods applied to ceria nanoislands. *J. Phys. Chem. Lett.* **2012**, *3*, 1956–1961. [CrossRef]
150. Nolan, M. Hybrid density functional theory description of oxygen vacancies in the CeO_2 (110) and (100) surfaces. *Chem. Phys. Lett.* **2010**, *499*, 126–130. [CrossRef]
151. Paier, J.; Penschke, C.; Sauer, J. Oxygen defects and surface chemistry of ceria: Quantum chemical studies compared to experiment. *Chem. Rev.* **2013**, *113*, 3949–3985. [CrossRef] [PubMed]
152. Preda, G.; Pacchioni, G. Formation of oxygen active species in Ag-modified CeO_2 catalyst for soot oxidation: A DFT study. *Catal. Today* **2011**, *177*, 31–38. [CrossRef]
153. Zhu, K.-J.; Liu, J.; Yang, Y.-J.; Xu, Y.-X.; Teng, B.-T.; Wen, X.-D.; Fan, M. A method to explore the quantitative interactions between metal and ceria from CO atoms in the M/CeO_2 catalysts. *Surf. Sci.* **2018**, *669*, 79–86. [CrossRef]
154. Tereshchuk, P.; Freire, R.L.H.; Ungureanu, C.G.; Seminovski, Y.; Kiejnac, A.; Da Silva, J.L.F. The role of charge transfer in the oxidation state change of Ce atoms in the TM13–CeO_2(111) systems (TM = Pd, Ag, Pt, Au): A DFT + U investigation. *Phys. Chem. Chem. Phys.* **2015**, *17*, 13520–13530. [CrossRef] [PubMed]
155. Piotrowski, M.J.; Tereshchuk, P.; Da Silva, J.L.F. Theoretical Investigation of Small Transition-Metal Clusters Supported on the CeO_2 (111) Surface. *J. Phys. Chem. C* **2014**, *118*, 21438–21446. [CrossRef]
156. Chen, L.-J.; Tang, Y.; Cui, L.; Ouyang, C.; Shi, S. Charge transfer and formation of Ce^{3+} upon adsorption of metal atom M (M = Cu, Ag, Au) on CeO_2 (100) surface. *J. Power Sources* **2013**, *234*, 69–81. [CrossRef]
157. Luches, P.; Pagliuca, F.; Valeri, S.; Illas, F.; Preda, G.; Pacchioni, G. Nature of Ag Islands and Nanoparticles on the CeO_2 (111) Surface. *J. Phys. Chem. C* **2012**, *116*, 1122–1132. [CrossRef]
158. Cui, L.; Tang, Y.; Zhang, H.; Hector, L.G., Jr.; Ouyang, C.; Shi, S.; Lib, H.; Chen, L. First-principles investigation of transition metal atom M (M = Cu, Ag, Au) adsorption on CeO_2 (110). *Phys. Chem. Chem. Phys.* **2012**, *14*, 1923–1933. [CrossRef] [PubMed]
159. Tang, Y.; Zhang, H.; Cui, L.; Ouyang, C.; Shi, S.; Tang, W.; Li, H.; Chen, L. Electronic states of metal (Cu, Ag, Au) atom on CeO_2 (111) surface: The role of local structural distortion. *J. Power Sources* **2012**, *197*, 28–37.
160. Branda, M.M.; Hernández, N.C.; Sanz, J.F.; Illas, F. Density functional theory study of the interaction of Cu, Ag, and Au atoms with the regular CeO_2 (111) surface. *J. Phys. Chem. C* **2010**, *114*, 1934–1941. [CrossRef]
161. Hay, P.J.; Martin, R.L.; Uddin, J.; Scuseria, G.E. Theoretical study of CeO_2 and Ce_2O_3 using a screened hybrid density functional. *J. Chem. Phys.* **2006**, *125*, 034712. [CrossRef] [PubMed]
162. Da Silva, J.L.F.; Ganduglia-Pirovano, M.V.; Sauer, J.; Bayer, V.; Kresse, G. Hybrid functionals applied to rare-earth oxides: The example of ceria. *Phys. Rev. B* **2007**, *75*, 045121. [CrossRef]
163. Yang, Z.; Yu, X.; Lu, Z.; Li, S.; Hermansson, K. Oxygen vacancy pairs on CeO_2 (110): A DFT + U study. *Phys. Lett. A* **2009**, *373*, 2786–2792. [CrossRef]
164. Dudarev, S.L.; Botton, G.A.; Savrasov, S.Y.; Humphreys, C.J.; Sutton, A.P. Electron-energy-loss spectra and the structural stability of nickel oxide: An LSDA+U study. *Phys. Rev. B* **1998**, *57*, 1505–1509. [CrossRef]
165. Branda, M.M.; Castellani, N.J.; Grau-Crespo, R.; de Leeuw, N.H.; Hernandez, N.C.; Sanz, J.F.; Neyman, K.M.; Illas, F.J. On the difficulties of present theoretical models to predict the oxidation state of atomic Au adsorbed on regular sites of CeO_2(111). *J. Chem. Phys.* **2009**, *131*, 094702.
166. Trovarelli, A.; Fornasiero, P. *Catalysis by Ceria and Related Materials*, 2nd ed.; Imperial College Press: London, UK, 2013.
167. Wen, X.-J.; Niu, C.-G.; Zhang, L.; Liang, C.; Zeng, G.-M. A novel Ag_2O/CeO_2 heterojunction photocatalysts for photocatalytic degradation of enrofloxacin: possible degradation pathways, mineralization activity and an in depth mechanism insight. *Appl. Catal. B* **2018**, *221*, 701–714. [CrossRef]
168. Xie, S.; Wang, Z.; Cheng, F.; Zhang, P.; Mai, W.; Tong, Y. Ceria and ceria-based nanostructured materials for photoenergy applications. *Nano Energy* **2017**, *34*, 313–337. [CrossRef]

169. Channei, D.; Inceesungvorn, B.; Wetchakun, N.; Ukritnukun, S.; Nattestad, A.; Chen, J.; Phanichphant, S. Photocatalytic degradation of methyl orange by CeO_2 and Fe–doped CeO_2 films under visible light irradiation. *Sci. Rep.* **2014**, *4*, 5757. [CrossRef] [PubMed]
170. Ren, H.; Koshy, P.; Chen, W.-F.; Qi, S.; Sorrell, C.C. Photocatalytic materials and technologies for air purification. *J. Hazard. Mater.* **2017**, *325*, 340–366. [CrossRef] [PubMed]
171. Li, Y.; Sun, Q.; Kong, M.; Shi, W.; Huang, J.; Tang, J.; Zhao, X. Coupling oxygen ion conduction to photocatalysis in mesoporous nanorod-like ceria significantly improves photocatalytic efficiency. *J. Phys. Chem. C* **2011**, *115*, 14050–14057. [CrossRef]
172. Sabari Arul, N.; Mangalaraj, D.; Whan Kim, T. Photocatalytic degradation mechanisms of self-assembled rose-flower-like CeO_2 hierarchical nanostructures. *Appl. Phys. Lett.* **2013**, *102*, 223115. [CrossRef]
173. Ko, J.W.; Kim, J.H.; Park, C.B. Synthesis of visible light-active CeO_2 sheets via mussel-inspired $CaCO_3$ mineralization. *J. Mater. Chem. A* **2013**, *1*, 241–245. [CrossRef]
174. Huang, Y.; Long, B.; Tang, M.; Rui, Z.; Balogun, M.-S.; Tong, Y.; Ji, H. Bifunctional catalytic material: An ultrastable and high-performance surface defect CeO_2 nanosheets for formaldehyde thermal oxidation and photocatalytic oxidation. *Appl. Catal. B* **2016**, *181*, 779–787. [CrossRef]
175. Xu, B.; Zhang, Q.; Yuan, S.; Zhang, M.; Ohno, T. Morphology control and photocatalytic characterization of yttrium-doped hedgehog-like CeO_2. *Appl. Catal. B* **2015**, *164*, 120–127. [CrossRef]
176. Ji, Y.; Ferronato, C.; Salvador, A.; Yang, X.; Chovelon, J.M. Degradation of ciprofloxacin and sulfamethoxazole by ferrous-activated persulfate: implications for remediation of groundwater contaminated by antibiotics. *Sci. Total Environ.* **2014**, *472*, 800–808. [CrossRef] [PubMed]
177. Huang, G.F.; Ma, Z.L.; Huang, W.Q.; Tian, Y.; Jiao, C.; Yang, Z.M.; Wan, Z.; Pan, A. Ag_3PO_4 Semiconductor Photocatalyst: Possibilities and Challenges. *J. Nanomater.* **2013**, *8*, 371356.
178. Yang, Z.-M.; Huang, G.-F.; Huang, W.-Q.; Wei, J.-M.; Yan, X.-G.; Liu, Y.-Y.; Jiao, C.; Wan, Z.; Pan, A. Novel Ag_3PO_4/CeO_2 composite with high efficiency and stability for photocatalytic applications. *J. Mater. Chem. A* **2014**, *2*, 1750–1756. [CrossRef]
179. Wu, C. Synthesis of Ag_2CO_3/CeO_2 microcomposite with visible light-driven photocatalytic activity. *Mater. Lett.* **2015**, *152*, 76–78. [CrossRef]
180. Wen, X.-J.; Niu, C.-G.; Huang, D.-W.; Zhang, L.; Liang, C.; Zeng, G.-M. Study of the photocatalytic degradation pathway of norfloxacin and mineralization activity using a novel ternary $Ag/AgCl$-CeO_2 photocatalyst. *J. Catal.* **2017**, *355*, 73–86. [CrossRef]
181. Linic, S.; Christopher, P.; Ingram, D.B. Plasmonic-metal nanostructures for efficient conversion of solar to chemical energy. *Nat. Mater.* **2011**, *10*, 911. [CrossRef] [PubMed]
182. Tanaka, A.; Hashimoto, K.; Kominami, H. Gold and copper nanoparticles supported on cerium(IV) oxide a photocatalyst mineralizing organic acids under red light irradiation. *ChemCatChem* **2011**, *3*, 1619–1623. [CrossRef]
183. Kominami, H.; Tanaka, A.; Hashimoto, K. Mineralization of organic acids in aqueous suspensions of gold nanoparticles supported on cerium(IV) oxide powder under visible light irradiation. *Chem. Commun.* **2010**, *46*, 1287–1289. [CrossRef] [PubMed]
184. Liu, I.-T.; Hon, M.-H.; Kuan, C.-Y.; Teoh, L.-G. Structure and optical properties of Ag/CeO_2 nanocomposites. *Appl. Phys. A* **2013**, *111*, 1181–1186. [CrossRef]
185. Liu, T.; Li, B.; Wang, Y.; Ge, Z.; Shi, J. Facile Synthesis of Ag/CeO_2 Mesoporous Composites with Enhanced Visible Light Photocatalytic Properties. *Asian J. Chem.* **2014**, *26*, 1355–1357.
186. Hao, Y.; Li, L.; Liu, D.; Yu, H.; Zhou, Q. The synergy of SPR effect and Z-scheme of Ag on enhanced photocatalytic performance of 3DOM Ag/CeO_2-ZrO_2 composite. *Mol. Catal.* **2018**, *447*, 37–46. [CrossRef]
187. Saravanan, R.; Agarwal, S.; Gupta, V.K.; Khan, M.M.; Gracia, F.; Mosquera, E.; Narayanan, V.; Stephen, A. Line defect Ce^{3+} induced $Ag/CeO_2/ZnO$ nanostructure for visible-light photocatalytic activity. *J. Photochem. Photobiol. A* **2018**, *353*, 499–506. [CrossRef]
188. Mittal, M.; Gupta, A.; Pandey, O.P. Role of oxygen vacancies in Ag/Au doped CeO_2 nanoparticles for fast photocatalysis. *Sol. Energy* **2018**, *165*, 206–216. [CrossRef]
189. Barsuk, D.; Zadick, A.; Chatenet, M.; Georgarakis, K.; Panagiotopoulos, N.T.; Champion, Y.; Moreira Jorge, A., Jr. Nanoporous silver for electrocatalysis application in alkaline fuel cells. *Mater. Des.* **2016**, *111*, 528–536. [CrossRef]

190. Lu, Q.; Rosen, J.; Zhou, Y.; Hutchings, G.S.; Kimmel, Y.C.; Chen, J.G.; Jiao, F. A selective and efficient electrocatalyst for carbon dioxide reduction. *Nat. Commun.* **2014**, *5*, 3242. [CrossRef] [PubMed]
191. Gonzalez-Macia, L.; Smyth, M.R.; Killard, A.J. Evaluation of a silver-based electrocatalyst for the determination of hydrogen peroxide formed via enzymatic oxidation. *Talanta* **2012**, *99*, 989–996. [CrossRef] [PubMed]
192. Sreeremya, T.S.; Krishnan, A.; Remani, K.C.; Patil, K.R.; Brougham, D.F.; Ghosh, S. Shape-selective oriented cerium oxide nanocrystals permit assessment of the effect of the exposed facets on catalytic activity and oxygen storage capacity. *ACS Appl. Mater. Interfaces* **2015**, *7*, 8545–8555. [CrossRef] [PubMed]
193. Meher, S.K.; Rao, G.R. Novel nanostructured CeO_2 as efficient catalyst for energy and environmental applications. *J. Chem. Sci.* **2014**, *126*, 361–372. [CrossRef]
194. Melchionna, M.; Fornasiero, P. The role of ceria-based nanostructured materials in energy applications. *Mater. Today* **2014**, *17*, 349–357. [CrossRef]
195. Li, G.; Zhang, X.; Wang, L.; Song, X.; Sun, Z. Promoting Effect of Au on the Nanoporous Ag/CeO_2 Composites Prepared by Dealloying for Borohydride Electro-Oxidation. *J. Electrochem. Soc.* **2013**, *160*, 1116–1122. [CrossRef]
196. Sun, S.; Xue, Y.; Wang, Q.; Li, S.; Huang, H.; Miao, H.; Liu, Z. Electrocatalytic activity of silver decorated ceria microspheres for the oxygen reduction reaction and their application in aluminium–air batteries. *Chem. Commun.* **2017**, *53*, 7921–7924. [CrossRef] [PubMed]

© 2018 by the authors. Licensee MDPI, Basel, Switzerland. This article is an open access article distributed under the terms and conditions of the Creative Commons Attribution (CC BY) license (http://creativecommons.org/licenses/by/4.0/).

Review

Molecular Orientations Change Reaction Kinetics and Mechanism: A Review on Catalytic Alcohol Oxidation in Gas Phase and Liquid Phase on Size-Controlled Pt Nanoparticles

Fudong Liu [1,2,†], **Hailiang Wang** [1,2,‡], **Andras Sapi** [1,2,§], **Hironori Tatsumi** [2,∥], **Danylo Zherebetskyy** [2], **Hui-Ling Han** [1,2], **Lindsay M. Carl** [1,2] **and Gabor A. Somorjai** [1,2,*]

1. Department of Chemistry, University of California, Berkeley, CA 94720, USA; lfd1982@gmail.com or fudong.liu@basf.com (F.L.); hailiang.wang@yale.edu (H.W.); sapia@chem.u-szeged.hu (A.S.); lyndahan@gmail.com (H.-L.H.); lindsaymishele@gmail.com (L.M.C.)
2. Materials Sciences Division, Lawrence Berkeley National Laboratory, Berkeley, CA 94720, USA; hironori_tatsumi@shokubai.co.jp (H.T.); zherebetskyy@gmail.com (D.Z.)
* Correspondence: somorjai@berkeley.edu; Tel.: +1-510-642-4053; Fax: +1-510-643-9668
† Current address: BASF Corporation, 25 Middlesex Essex Turnpike, Iselin, NJ 08830, USA
‡ Current address: Department of Chemistry, Yale University, 300 Heffernan Dr, West Haven, CT 06516, USA
§ Current address: Department of Applied and Environmental Chemistry, University of Szeged, Rerrich Square 1, H-6720 Szeged, Hungary
∥ Current address: Nippon Shokubai Co., Ltd., 992-1 Aza-Nishioki, Okihama, Aboshi-ku, Himeji, Hyogo 671-1282, Japan

Received: 30 April 2018; Accepted: 26 May 2018; Published: 27 May 2018

Abstract: Catalytic oxidation of alcohols is an essential process for energy conversion, production of fine chemicals and pharmaceutical intermediates. Although it has been broadly utilized in industry, the basic understanding for catalytic alcohol oxidations at a molecular level, especially under both gas and liquid phases, is still lacking. In this paper, we systematically summarized our work on catalytic alcohol oxidation over size-controlled Pt nanoparticles. The studied alcohols included methanol, ethanol, 1-propanol, 2-propanol, and 2-butanol. The turnover rates of different alcohols on Pt nanoparticles and also the apparent activation energy in gas and liquid phase reactions were compared. The Pt nanoparticle size dependence of reaction rates and product selectivity was also carefully examined. Water showed very distinct effects for gas and liquid phase alcohol oxidations, either as an inhibitor or as a promoter depending on alcohol type and reaction phase. A deep understanding of different alcohol molecular orientations on Pt surface in gas and liquid phase reactions was established using sum-frequency generation spectroscopy analysis for in situ alcohol oxidations, as well as density functional theory calculation. This approach can not only explain the entirely different behaviors of alcohol oxidations in gas and liquid phases, but can also provide guidance for future catalyst/process design.

Keywords: catalytic alcohol oxidation; gas phase; liquid phase; Pt nanoparticles; sum-frequency generation spectroscopy; surface molecular orientation; density functional theory calculation

1. Introduction

Catalytic partial oxidation and complete oxidation of alcohols over platinum group metals (PGM) or metal oxide catalysts are fundamental processes not only in energy conversion, such as in fuel cells [1,2], but also in fine chemical synthesis and the pharmaceutical industry [3–7]. Usually, the production of aldehydes and ketones is performed through alcohol oxidation in the gas phase at high temperatures, while the production of fine chemicals and pharmaceutical intermediates is performed in the liquid phase at low temperatures [8,9]. Many researchers have focused on the synthesis of novel, highly efficient, poisoning-resistant, or low-cost catalysts to improve productivity and selectivity, as well as to lower environmental impact [6,10–18]. However, very few studies have focused on the systematic comparison on gas phase and liquid phase alcohol oxidations over the same PGM or metal oxide catalysts at the molecular level, which is very important for the basic understanding of the reaction kinetics and mechanisms to advance and improve the catalyst and process designs for practical application.

This review paper systematically summarizes our previous work in the catalytic alcohol oxidation area, in both gas phase and liquid phase over size-controlled Pt nanoparticles [9,19–22]. The studied alcohols included C1–C4 molecules, i.e., methanol (MeOH), ethanol (EtOH), 1-propanol (1-PrOH), 2-propanol (2-PrOH), and 2-butanol (2-BuOH). Detailed comparisons of the reaction rates in both phases and the Pt nanoparticle size dependence of reaction rates, as well as product selectivity, the apparent activation energy of alcohol oxidations in both phases, and also the response to co-existing water under different reaction conditions, are all included herein. To understand the intrinsic reasons at the molecular level for differences in reaction kinetics and mechanisms in alcohol oxidation under gas and liquid phases, the sum-frequency generation (SFG) vibrational spectroscopy measurements on Pt surface under reaction conditions were conducted and discussed in detail. In aid of density functional theory (DFT) computational modeling, different alcohol molecular orientations/configurations on Pt surface in the gas phase and liquid phase reactions were confirmed, which well explained the phenomena that were observed with striking differences.

2. Results and Discussion

2.1. Turnover Rate Comparison for Alcohol Oxidation in Gas Phase and Liquid Phase

Figure 1 shows the turnover frequency (TOF) of different alcohols in catalytic oxidation reactions producing carbonyl compounds in both gas phase and liquid phase. As we can observe, different alcohols in the gas phase oxidation reaction showed distinct turnover rates; for example, at 60 °C, MeOH showed the highest TOF, followed by EtOH, 2-PrOH and 1-PrOH. The saturated vapor pressure of MeOH, EtOH, 1-PrOH and 2-PrOH at 20 and 60 °C, either cited from literature or calculated using Antoine Equation, are also shown here [23–25]. From Figure 1, it can also be seen that, interestingly, there seems to be a good correlation between gas phase alcohol oxidation reaction rates and alcohol vapor pressure. Alcohols with higher vapor pressure such as MeOH, EtOH, and 2-PrOH have many more dynamic molecules in the gas phase; thus, it can reach the catalyst surface, react to form intermediates/products, and leave the catalyst surface more efficiently. In contrast, alcohol with lower vapor pressure such as 1-PrOH has less dynamic molecules in the gas phase, and these molecules are either "reluctant" to reach the catalyst surface or "stick" to the surface upon contact without leaving quickly, thus resulting in the lower reaction rates in the gas phase oxidation reaction. For 2-BuOH oxidation in gas phase, the reaction was carried out at 80 °C; therefore, the direct comparison of reaction rates between 2-BuOH and other alcohols was not performed here.

For the liquid phase oxidation reactions using pure alcohols, in most cases, such as for MeOH, EtOH, 2-PrOH and 2-BuOH, the turnover rates were lower than those in the gas phase reaction. 1-PrOH was an exception that the liquid phase reaction rate under such condition was higher than that in the gas phase. For gas phase alcohol oxidations, it should be noted that the alcohol to oxygen ratio was controlled at 1:5 (~0.48 mM of alcohols and ~2.41 mM of O_2), while in the liquid phase reaction this alcohol to oxygen ratio was much higher (~4 orders of magnitude depending on alcohol density) than that in the gas phase due to much higher density of alcohols in pure liquid phase. Therefore, for reasonable comparison, we diluted the liquid phase alcohols to one thousandth using a neutral solvent, heptane, which does not show a clear impact on the reaction kinetics of alcohol oxidations under similar reaction conditions [22]. In this way, the liquid phase alcohol concentrations ranged from 10 to 24 mM, and the dissolved O_2 concentration in the liquid phase (alcohol plus heptane) was about 16.7 mM, making the liquid phase reaction conditions much more similar/comparable to the gas phase reaction conditions. It is evident that, even under comparable alcohol molecular density on Pt nanoparticle surface after 1000 times dilution including MeOH, EtOH, 1-PrOH and 2-PrOH, the reaction rates in the liquid phase were about 1~4 magnitude slower than those in the gas phase reaction. The dilution of 2-BuOH in the liquid phase was not performed, but based on the dilution results for other alcohols, the reaction rate of 2-BuOH would be further decreased upon dilution resulting in much lower activity. These results suggest that the reaction rates of catalytic alcohol oxidation heavily depended on the reaction phase (gas phase versus liquid phase), and the intrinsic root cause for such discrepancy should be understood at the molecular level.

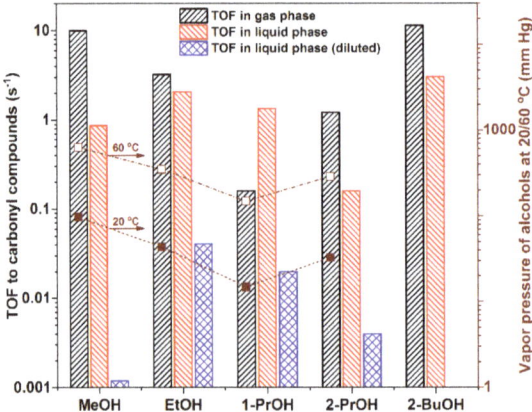

Figure 1. Turnover rates of alcohol oxidation to carbonyl compounds in gas and liquid phases over 6 nm Pt/MCF-17 (4.5 nm Pt/MCF-17 for 1-PrOH oxidation). Gas phase reaction: 1.33 kPa alcohol, 6.67 kPa O_2, 94.66 kPa He, 60 °C reaction temperature for MeOH, EtOH, 1-PrOH, 2-PrOH and 80 °C reaction temperature for 2-BuOH. Liquid phase reaction: 15 mL alcohol, dissolved oxygen under 100 kPa for MeOH, EtOH, 1-PrOH, 2-PrOH (60 °C reaction temperature) and 300 kPa for 2-BuOH (80 °C reaction temperature). Liquid phase reaction (1000 times diluted): 15 mL heptane, 15 µL alcohol with dissolved oxygen under 100 kPa for MeOH, EtOH, 1-PrOH, 2-PrOH at 60 °C reaction temperature. The vapor pressure of MeOH, EtOH, 1-PrOH, 2-PrOH at 20 and 60 °C is also presented herein. (TOF data for MeOH, EtOH, 1-PrOH, 2-PrOH, 2-BuOH oxidations were reported in [9,19–22], respectively).

2.2. Size Effect of Pt Nanoparticles on Alcohol Oxidation in Gas Phase and Liquid Phase

Both gas phase and liquid phase alcohol oxidations were carried out on Pt nanoparticles with precisely controlled particle sizes, i.e., 2–9 nm Pt loaded into MCF-17 mesoporous silica. Accordingly, we could study the Pt nanoparticle size dependence of the turnover rates, as well as the product selectivity for different alcohols under both reaction conditions.

As shown in Figure 2A, for all the alcohol oxidations in the gas phase, including MeOH, EtOH, 1-PrOH, 2-PrOH at 60 °C and 2-BuOH at 80 °C, the turnover rates all showed a monotonic increase as the Pt nanoparticle size grew (except a single point for MeOH oxidation on 8 nm Pt). A very similar trend was also observed for all alcohol oxidations in the liquid phase, as shown in Figure 2B. These results indicate that the alcohol oxidation reactions preferentially took place on step sites or terrace sites on larger Pt nanoparticles, while the corner sites or edge sites on smaller Pt nanoparticle were not favorable for alcohol oxidations, probably due to their too-strong affinity to oxygenated species blocking the catalyst surface, which was not beneficial to the rate-determining dehydrogenation process of alcohol adsorbates [1,26].

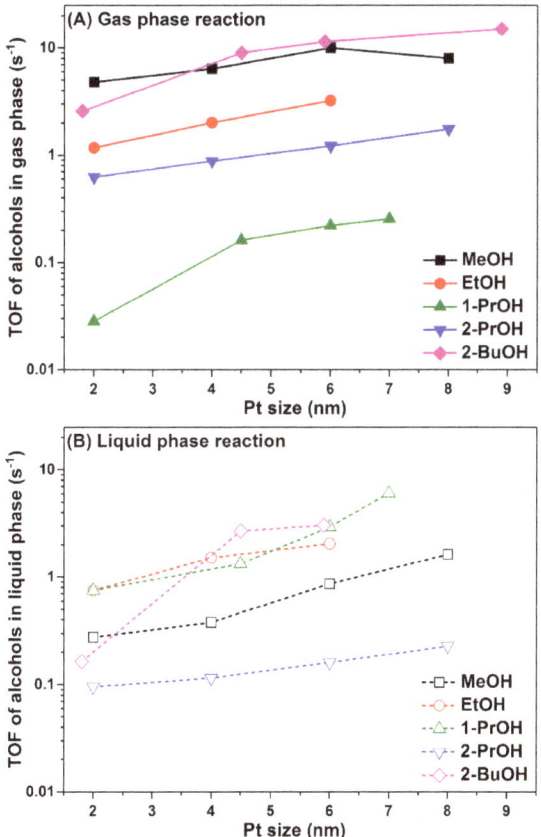

Figure 2. Size effect of Pt nanoparticles on TOF values of MeOH, EtOH, 1-PrOH, 2-PrOH oxidation at 60 °C and 2-BuOH oxidation at 80 °C. (**A**) Gas phase reaction: 1.33 kPa alcohol, 6.67 kPa O_2, 94.66 kPa He; (**B**) Liquid phase reaction: 15 mL alcohol, dissolved oxygen under 100 kPa (300 kPa for 2-BuOH). (TOF data for MeOH, EtOH, 1-PrOH, 2-PrOH, 2-BuOH oxidations as a function of Pt nanoparticle sizes were reported in [9,19–22], respectively).

For all alcohol oxidations that we studied in both gas phase and liquid phase, except for CO_2 resulting from complete oxidation, the products were mainly carbonyl compounds from partial oxidation. Figure 3A shows the selectivity to carbonyl compounds in the gas phase alcohol oxidations. For gas phase MeOH oxidation, the main products were formaldehyde (less) and methyl formate (more), and the selectivity to these two compounds was about 60–70%. No clear correlation between formaldehyde plus methyl formate selectivity and Pt nanoparticle size was observed, except that the highest selectivity was observed on 4–6 nm Pt nanoparticles. For gas phase EtOH oxidation, the main product was acetaldehyde, with selectivity as high as 97%. For gas phase 1-PrOH oxidation, the main product was propanal, and similar to the MeOH case, the highest selectivity to propanal (>80%) was also observed on 4–6 nm Pt nanoparticles. For gas phase 2-PrOH oxidation, acetone was the only product. Moreover, for gas phase 2-BuOH oxidation, the selectivity to butanone on 4–6 nm Pt nanoparticles (ca. 97%) was also slightly higher than that on 2 nm Pt. Figure 3B shows the selectivity to carbonyl compounds in liquid phase alcohol oxidations. For liquid phase MeOH oxidation, interestingly, much more formaldehyde was produced than methyl formate, and the selectivity to formaldehyde plus methyl formate (ca. 80–90%) was also much higher than that in the gas phase reaction (60–70%). Smaller Pt nanoparticles (such as 2 nm) were more likely to catalyze the deep oxidation of MeOH, thus resulting in the formation of more methyl formate, while larger Pt nanoparticles (such as 4–8 nm) were more favorable for formaldehyde formation with monotonic correlation with particle size. For all other alcohols, including EtOH, 1-PrOH, 2-PrOH and 2-BuOH, the selectivity to carbonyl compounds in the liquid phase oxidation reactions were either similar or higher than those in the gas phase reactions (see 1-PrOH data for more apparent comparison), implying that the complete oxidation of alcohols in liquid phase was actually inhibited to a certain extent, probably due to the different molecular density or molecular orientation on the Pt nanoparticle surface.

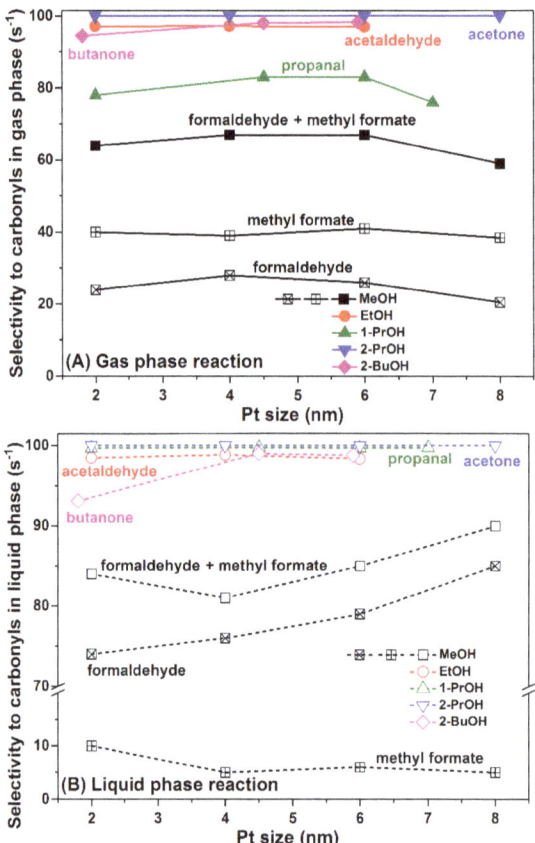

Figure 3. Size effect of Pt nanoparticles on product selectivity to carbonyl compounds (other than CO_2) of MeOH, EtOH, 1-PrOH, 2-PrOH oxidation at 60 °C and 2-BuOH oxidation at 80 °C. (**A**) Gas phase reaction: 1.33 kPa alcohol, 6.67 kPa O_2, 94.66 kPa He; (**B**) Liquid phase reaction: 15 mL alcohol, dissolved oxygen under 100 kPa for MeOH, EtOH, 1-PrOH, 2-PrOH, and 300 kPa for 2-BuOH. (Selectivity data for MeOH, EtOH, 1-PrOH, 2-PrOH, 2-BuOH oxidations as a function of Pt nanoparticle sizes were reported in [9,19–22], respectively).

2.3. Different Activation Energies of Alcohol Oxidation in Gas Phase and Liquid Phase on Pt Nanoparticles

To further investigate the difference of reaction kinetics for alcohol oxidations in the gas phase and liquid phase, the apparent activation energy (Ea) on 4 nm Pt/MCF-17 for most alcohol oxidations

(4.5 nm Pt/MCF-17 for 1-PrOH oxidation) was measured and presented in Figure 4. It is very interesting to see that the apparent activation energy for all alcohol oxidations in the gas phase was much higher than that in the liquid phase, although under such reaction conditions the gas phase turnover rates were much higher than those in the liquid phase. This means that the gas phase alcohol oxidations are more sensitive to the reaction temperature, while the liquid phase alcohol oxidations do not. In practical application, if it is preferable to conduct the alcohol oxidations at higher operation temperatures, gas phase reactions are highly recommended, while if it is preferable to conduct the alcohol oxidations at lower operation temperatures, the liquid phase reactions are probably more suitable. However, the oxygen mass transfer in the liquid phase is much slower than that in the gas phase (e.g., regarding to oxygen diffusion coefficient D_{O2}, $D_{O2 \text{ in water, 283 K}}$: 1.54×10^{-5} cm^2/s, $D_{O2 \text{ in N2, 1 atm, 273 K}}$: 0.181 cm^2/s) [27,28]. It is necessary to improve the oxygen diffusion capacity in order to increase the total product yields in alcohol oxidations in the liquid phase. Besides the oxygen diffusion difference between gas phase and liquid phase reactions, the distinct alcohol molecular orientations on Pt surface in two different phases might be another important reason for activation energy discrepancy, and will be discussed in detail in the SFG spectra analysis and DFT calculation sections.

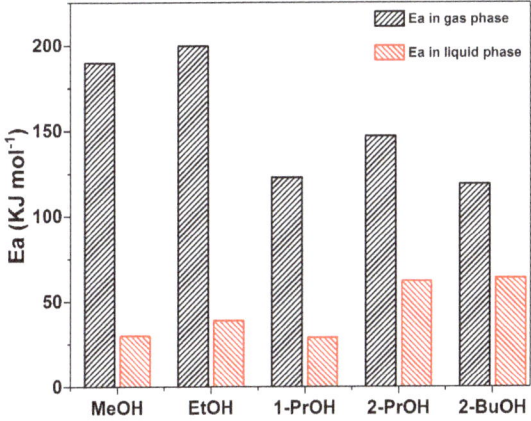

Figure 4. Apparent activation energy (Ea) of alcohol oxidations in gas and liquid phases over 4 nm Pt/MCF-17 (4.5 nm Pt/MCF-17 for 1-PrOH oxidation). Gas phase reaction: 1.33 kPa alcohol, 6.67 kPa O_2, 94.66 kPa He; Liquid phase reaction: 15 mL alcohol, dissolved oxygen under 100 kPa for MeOH, EtOH, 1-PrOH, 2-PrOH, and 300 kPa for 2-BuOH. (Ea data for MeOH, EtOH, 1-PrOH, 2-PrOH, 2-BuOH oxidations were reported in [9,19–22], respectively).

2.4. H_2O Effect on Alcohol Oxidation in Gas Phase and Liquid Phase on Pt Nanoparticles

H_2O is one of the products of the complete or partial oxidation of alcohols, especially in gas phase reactions, where the selectivity to carbonyl compounds is not as high as that in the liquid phase reactions. Therefore, it is indispensable to check the H_2O effect on alcohol oxidation not only in the gas phase but also in the liquid phase, which is quite essential for practical application.

As the results of relative turnover rates shown in Figure 5 demonstrate, for the gas phase MeOH, EtOH, 1-PrOH and 2-PrOH oxidations, water vapor definitely inhibited the reaction rates significantly, with the TOF values dramatically increasing upon water vapor addition. This could be simply explained by the competitive adsorption of H_2O onto the Pt surface, thus obviously blocking the active sites for catalytic alcohol oxidations. However, in the case of gas phase 2-BuOH oxidation, the water vapor addition showed some promotion effect at medium H_2O doping amounts (i.e., H_2O content of 0.17 and 0.33), which seemed to be unusual.

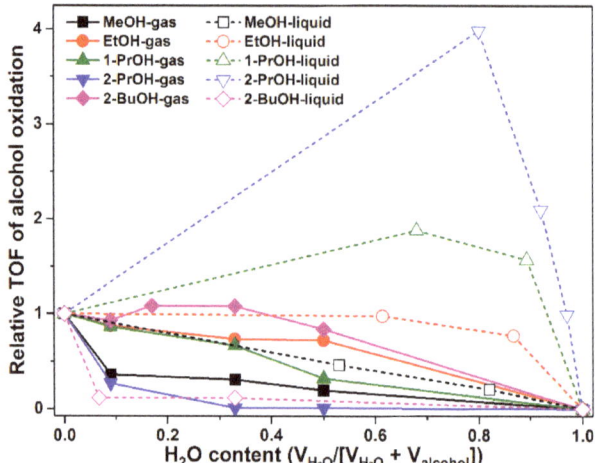

Figure 5. Effect of H$_2$O addition on relative TOF of alcohol oxidations over Pt/MCF-17 in gas and liquid phases. Gas phase reaction: 1.33 kPa alcohol, 0.13–1.33 kPa water vapor, 6.67 kPa O$_2$, He balance (in total 102.66 kPa); Liquid phase reaction: 5–10 mL alcohol, 0–10 mL distilled water, in total 15 mL volume, dissolved oxygen under 100 kPa for MeOH, EtOH, 1-PrOH, 2-PrOH and 300 kPa for 2-BuOH. Catalysts and reaction temperatures for both gas and liquid phases: 4 nm Pt/MCF-17 at 60 °C for MeOH, EtOH, 2-PrOH, 4.5 nm Pt/MCF-17 at 60 °C for 1-PrOH, and 6 nm Pt/MCF-17 at 80 °C for 2-BuOH. All data were normalized by TOF value without water addition. (H$_2$O effect data for MeOH, EtOH, 1-PrOH, 2-PrOH, 2-BuOH oxidations were reported in [9,19–22], respectively).

In the case of the liquid phase alcohol oxidations, the responses to aqueous water addition were totally different from case to case. For example, with regard to liquid MeOH oxidation, H$_2$O showed a nearly linear inhibition effect on reaction rate, but still the inhibition effect was not as strong as that in the gas phase reaction. In the case of the liquid phase EtOH oxidation, the inhibition effect of H$_2$O seemed to be mitigated to a certain extent. While, for the liquid phase 1-PrOH and 2-PrOH oxidations, H$_2$O actually acted as a "promoter" or "co-catalyst", which significantly increased the turnover rates. Such striking difference of reaction rates in response to aqueous water in the liquid phase alcohol oxidations comparing to response to water vapor in the gas phase alcohol oxidations was mainly due to the totally different alcohol molecular density and/or alcohol molecular orientation on the Pt surface. As for the liquid phase 2-BuOH oxidation, the impact of aqueous water on reaction rate was totally opposite to other alcohols. Even with a very small amount of aqueous water addition, such as an H$_2$O content of 0.07, the turnover rate dramatically decreased to ca. 12% of the initial value, indicating that aqueous water here actually acted as a "poisoning agent" for the liquid phase 2-BuOH oxidation.

So far, totally opposite effects were observed for H$_2$O on gas phase and liquid phase 2-BuOH oxidations, in contrast to other alcohols, which can probably be explained by the hydrophilicity difference of alcohols. MeOH, EtOH, 1-PrOH and 2-PrOH are all miscible in water, while 2-BuOH has a solubility of only 12.5 g per 100 mL of H$_2$O due to the existence of more hydrophobic alkyl chains [9,29]. The capping agent that we used for Pt nanoparticle synthesis, which was polyvinylpyrrolidone (PVP), actually showed amphiphilicity. In the case of the gas phase 2-BuOH oxidation with relatively high mobility of alcohol and H$_2$O molecules, once the 2-BuOH molecules had reached and attached to the Pt surface, H$_2$O could not be adsorbed onto the surface anymore in any significant amount due to the hydrophobic nature of the 2-BuOH molecules. Therefore, water vapor only showed a slight inhibition effect, or even some promotion effect (probably due to the more hydroxyl group formation in the presence of H$_2$O) [30,31], on gas phase 2-BuOH oxidation. Contrastingly, in the case of the liquid phase 2-BuOH oxidation with relatively low mobility of alcohol and H$_2$O molecules, aqueous water would preferably gather around the Pt surface due to the hydrophobic nature of 2-BuOH. Such aqueous water layer blocked the access to the surface active sites thus resulting in the decrease of turnover rate in the liquid 2-BuOH oxidation [9].

2.5. Case Study of 1-PrOH Oxidation Using SFG Spectra Analysis on Pt Thin Film and DFT Calculation in Gas and Liquid Phases

To fully understand the picture of how alcohol molecules interact with the Pt surface under different reaction conditions, taking 1-PrOH as first example, we conducted SFG spectra study at 60 °C, which is an in situ technique with surface-specific characteristics, on Pt thin films prepared by electron-beam deposition. Figure 6 shows the SFG spectra of 1-PrOH in gas phase on Pt thin film during reaction at 60 °C with 101.33 kPa of O$_2$ and different partial pressures of gas phase 1-PrOH, as well as the SFG spectra of 1-PrOH on Pt thin films at 60 °C purged by N$_2$ in the gas phase and liquid phase. As can clearly be seen from Figure 6a,b, the SFG peaks that can be assigned to symmetric CH$_2$ stretching mode at ca. 2840 cm^{-1}, symmetric CH$_3$ stretching mode at ca. 2870 cm^{-1}, asymmetric CH$_2$ stretching mode at ca. 2910 cm^{-1}, CH$_3$ Fermi resonance at ca. 2935 cm^{-1}, and asymmetric CH$_3$ stretching mode at ca. 2970 cm^{-1} can be observed on the Pt surface under 1.87 and 9.07 kPa of 1-PrOH with O$_2$. However, these spectra showed noticeable differences not only in the strength of CH$_2$ peaks but also in the ratios between asymmetric and symmetric methyl stretches. This is absolutely clear evidence that surface 1-PrOH molecule orientation on Pt is highly dependent on the alcohol molecular density in the gas phase. It should be noted that our SFG spectra were measured under *ppp* polarization. Therefore, the absolute 1-PrOH molecule orientation cannot be directly determined. However, SFG theory predicts that a change in the orientation of specific functional groups (such as –CH$_3$ groups in this study) relative to the studied surface can result in the intensity ratio change of different vibration modes [22]. In such studies, the surface of Pt was considered to possess C$_{\infty v}$ symmetry, while the 1-PrOH molecule orientation on the Pt surface was assumed to be isotropic with regard to the azimuthal angle to the z-axis. Therefore, the average tilt angle of –CH$_3$ group from the Pt surface (θ) could be described by such a measurement, and changes in the asymmetric/symmetric mode ratio among the spectra were accordingly representative of a change of θ [21]. A low value of θ describes a molecule with its methyl group pointing up from the surface ("standing up" configuration), and a high value describes a molecule close to the surface ("lying down" configuration) [22]. The ratio of asymmetric/symmetric stretches of –CH$_3$ group under 1.87 kPa of 1-PrOH with O$_2$ was ca. 0.5:1, while this ratio significantly increased to 2:1 under 9.07 kPa of 1-PrOH with O$_2$, which was four times higher. This indicates a significant change in θ between low and high 1-PrOH partial pressure, and thus a different molecular orientation on the Pt surface.

Furthermore, as shown in Figure 6c,d, we also measured the SFG spectra of 1-PrOH on the Pt surface under N$_2$ purge with 10.67 kPa partial pressure in the gas phase and pure 1-PrOH in the liquid phase. The SFG spectrum recorded for gas phase 1-PrOH under such conditions showed peaks that could be assigned to –CH$_3$ groups with symmetric stretching mode at ca. 2870 cm^{-1}, strong Fermi resonance at ca. 2935 cm^{-1}, and asymmetric stretching mode at ca. 2960 cm^{-1}, as well as –CH$_2$ groups as weak shoulders with symmetric stretching mode at ca. 2840 cm^{-1} and asymmetric stretching mode at ca. 2910 cm^{-1}. This spectrum was pretty similar to the one recorded under O$_2$ with 1-PrOH with a relatively larger partial pressure in Figure 6b, although in this case, both the asymmetric and symmetric stretches from –CH$_3$ and –CH$_2$ groups showed some increase in peak intensity, mainly due to the higher 1-PrOH density on Pt surface. Contrastingly, the SFG spectrum recorded for liquid phase 1-PrOH showed significantly changed peak patterns compared to the gas phase, with slightly decreased peak intensity in –CH$_2$ stretching modes and greatly increased intensity ratio between asymmetric and symmetric stretching modes from –CH$_3$. We believe that the average tilt angle of –CH$_3$ group from Pt surface, θ, for 1-PrOH in the liquid phase became much smaller than that in the gas phase, which means that the molecular structure in the liquid phase was more ordered and more preferentially in a "standing up" configuration than the "lying down" configuration in the gas phase [21].

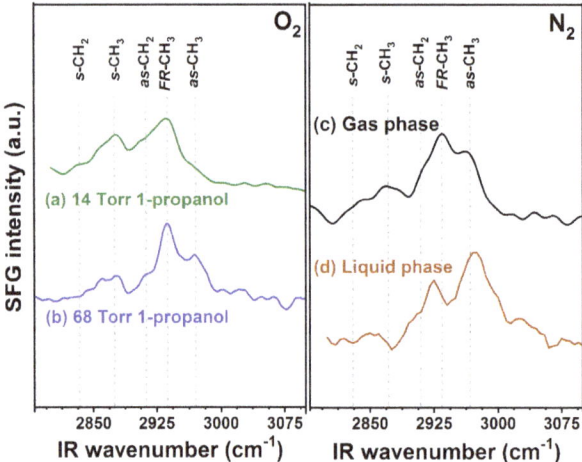

Figure 6. SFG spectra collected for gas phase 1-PrOH on Pt thin film during reactions at 60 °C with 101.33 kPa of O_2: (**a**) 1.87 kPa (14 Torr) of 1-PrOH; (**b**) 9.07 kPa (68 Torr) of 1-PrOH. SFG spectra collected for 1-PrOH on Pt thin film at 60 °C purged with N_2: (**c**) gas phase, 10.67 kPa of 1-PrOH; (**d**) liquid phase 1-PrOH. (SFG data for 1-PrOH oxidation were reported in [21]. Reproduced with permission from [21]. Copyright 2018 American Chemical Society.).

To better understand the molecular orientation of 1-PrOH on Pt surface in gas and liquid phases, we performed DFT theoretical calculation to simulate the concentration-dependent 1-PrOH configurations on Pt(111), which is the dominant surface for Pt nanoparticles used for alcohol oxidation reactions. More details about DFT calculation, as well as comprehensive results, can be found in our previous publication [21], while Figure 7 herein shows the minimum energy configurations of 1-PrOH molecules on Pt(111) surface with a surface molecular coverage of 0.94 molecules/nm^2, which represents the gas phase condition, as well as a surface molecular coverage of 3.75 molecules/nm^2, which represents the liquid phase condition. As we can see, under the gas phase condition, the 1-PrOH molecules were nearly "lying down" on the Pt surface, with the bisectrix connecting hydroxyl-O and methyl-C forming 6° angle relative to the surface (as shown in Table 1). Under the liquid phase condition, the 1-PrOH molecules were nearly "standing up" on the Pt surface with the bisectrix forming 41° angle relative to the surface (as shown in Table 1). These results are very consistent with the SFG spectral data and well explain the relative peak intensity changes that we observed for 1-PrOH on Pt thin film in the gas phase versus liquid phase. It should also be noted that, as shown in Table 1, the distance between the hydrogen atoms from alcohol hydroxyl group in 1-PrOH and Pt surface under the liquid phase condition was calculated as 0.261 nm, which was 0.056 nm closer to the Pt surface than the corresponding distance under the gas phase condition, i.e., 0.317 nm. This indicates that the hydroxyl group, and possibly also the α-H connecting to the same carbon atom in the 1-PrOH molecule, were much more easily activated/dehydrogenated in the liquid phase than in the gas phase, well explaining why the activation energy for 1-PrOH oxidation in the liquid phase was much lower than that in the gas phase.

Table 1. 1-Propanol molecule orientation as angle of C–C bonds relative to surface normal, and nearest surface–molecule distance for different concentrations of molecules on Pt(111) surface [21].

Concentration (molecules/nm^2)	α (°)	H$_{alc}$-Pt (nm)
0.94	6	0.317
3.75	41	0.261

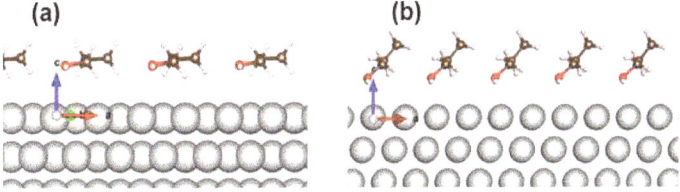

Figure 7. Minimum energy configurations of 1-PrOH molecules on Pt(111) surface for (**a**) gas phase (0.94 molecules/nm^2) and (**b**) liquid phase (3.75 molecules/nm^2) from DFT calculation (Pt—gray, C—brown, O—red, H—pink). (DFT results for 1-PrOH oxidation were reported in [21]. Reprinted with permission from [21]. Copyright 2018 American Chemical Society.).

2.6. Case Study of 2-PrOH Oxidation Using SFG Spectra Analysis on Pt Nanoparticles and DFT Calculation in Gas Liquid Phases

Similar to the 1-PrOH case, we also performed SFG spectra analysis of 2-PrOH oxidation in both gas phase and liquid phase on 4 nm Pt nanoparticles, which was more representatively reflective of the real catalytic reactions that we studied. As shown in Figure 8a, in the gas phase 2-PrOH oxidation reaction, three noticeable SFG peaks showed up, which could be ascribed to symmetric stretches of CH$_{3,ss}$ at ca. 2875 cm^{-1}, Fermi resonance mode of CH$_{3,fr}$ at ca. 2940 cm^{-1}, and asymmetric stretches of CH$_{3,as}$ at ca. 2970 cm^{-1}. The intensity ratio between asymmetric and symmetric stretches of –CH$_3$ was relatively small in this case. Not surprisingly, as shown in Figure 8b, in the liquid phase 2-PrOH oxidation reaction, all three SFG peaks ascribed to CH$_{3,ss}$, CH$_{3,fr}$, CH$_{3,as}$ showed up, along with a stretching peak from –CH group at ca. 2905 cm^{-1}; however, the intensity ratio between asymmetric and symmetric stretches of –CH$_3$ showed significant increase compared to the gas phase spectrum. These results indicate that the average tilt angle of –CH$_3$ group from Pt nanoparticle surface, θ, must have changed from a high value in the gas phase to a low value in the liquid phase, suggesting a change in the 2-PrOH molecular configurations from "lying down" to "standing up" on Pt surface, respectively. More detailed discussions about the possibility of 2-PrOH orientation varieties in different phases determined by SFG experimental data can be found in our previous publication [22].

To further confirm the 2-PrOH molecular configuration on the Pt surface, similarly to the 1-PrOH case, DFT theoretical calculation was also performed in our study. As the nanoparticles and nanoparticle-molecule complexes were too large for ab initio calculation, we still used the most dominant Pt(111) surface to investigate the concentration/phase dependence of 2-PrOH molecular orientations. As shown in Figure 9a, the minimum energy configuration of 2-PrOH molecules on Pt(111) surface with a low surface coverage of 0.94 molecules/nm^2, which is representative of the gas phase condition, showed a "lying down" pattern, with both of the C–C bonds forming 86° relative to the surface normal (Table 2). In contrast, as shown in Figure 9b, the minimum energy configuration of 2-PrOH molecules on the Pt(111) surface with high surface coverage of 3.75 molecules/nm^2, which is representative of the liquid phase condition, showed a "standing up" pattern with one C–C bond forming 84° and the other forming 38° relative to the surface normal (Table 2). The steric molecular interaction effect can easily explain this phenomenon, i.e., alcohol molecules are forced to take the "standing up" position in order to pack more molecules on the Pt surface [22]. As shown in Table 2, the 2-PrOH molecular orientation change from "lying down" to "standing up" on the Pt surface also resulted in the distance change between α-H and the nearest Pt atom, i.e., 0.445 nm in the gas phase and 0.257 nm in the liquid phase. Obviously, in such cases, much easier cleavage/dehydrogenation of α-H from α-C (which was considered as the rate-determining step in 2-PrOH oxidation [1]) can be achieved in the liquid phase than in the gas phase. This is also consistent with the observation that 2-PrOH oxidation showed much lower activation energy in the liquid phase than that in the gas phase [22].

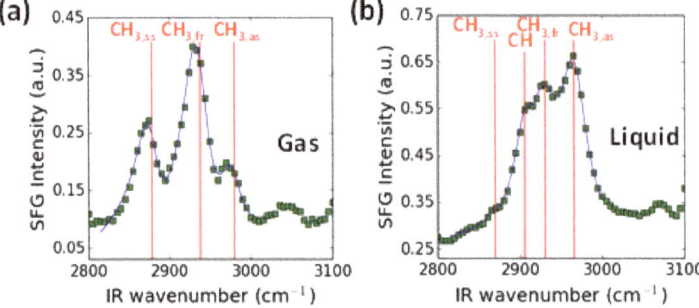

Figure 8. SFG spectra obtained on the surface of 4 nm Pt nanoparticles for 2-PrOH oxidation under (**a**) gas phase and (**b**) liquid phase reaction conditions. (SFG data for 2-PrOH oxidation were reported in [22]. Reprinted with permission from [22]. Copyright 2014 American Chemical Society.).

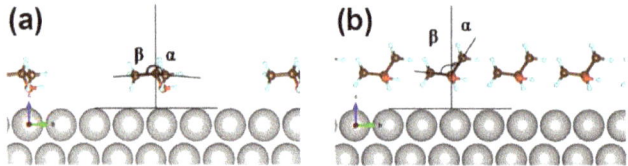

Figure 9. Minimum energy configurations of 2-PrOH molecules on Pt(111) surface for (**a**) gas phase (0.94 molecules/nm^2) and (**b**) liquid phase (3.75 molecules/nm^2) from DFT calculation (Pt—gray, C—brown, O—red, H—blue). (DFT results for 2-PrOH oxidation were reported in [22]. Reprinted with permission from [22]. Copyright 2014 American Chemical Society.).

Table 2. 2-Propanol molecular orientations as angles of C–C bonds relative to surface normal, and nearest α-H-Pt distances for different concentrations of 2-propanol molecules on Pt(111) surface [22].

Concentration (molecules/nm$_2$)	α (°)	β (°)	α-H-Pt (nm)
0.94	86	86	0.445
3.75	38	84	0.257

2.7. Case Study of 2-BuOH Oxidation Using SFG Spectra Analysis on Pt Thin Film in Gas Phase: O_2 and H_2O Effect

To better understand the 2-BuOH oxidation reaction in the gas phase, we performed SFG spectra analysis using various oxygen concentrations as well as added water vapor to see if there was any influence on the 2-BuOH molecular orientation. As shown in Figure 10a, we first collected the SFG spectrum under 2-BuOH with N_2, and the peaks ascribed to –CH_3 symmetric stretching mode at ca. 2880 cm^{-1}, –CH_3 Fermi resonance mode at ca. 2945 cm^{-1}, and –CH_3 asymmetric stretching mode at ca. 2975 cm^{-1} could be observed. Based on the SFG results for 1-PrOH, 2-PrOH on the Pt surface, and judging from the intensity ratio between asymmetric and symmetric stretching mode for 2-BuOH case, it can be empirically deduced that the 2-BuOH molecules also had a "lying down" configuration on the Pt surface in the gas phase. Interestingly, in the presence of oxygen, even at a relatively low level (O_2:2-BuOH = 3:1), a much higher SFG signal for 2-BuOH was observed than that in the presence of inert N_2. Further increasing the oxygen content to O_2:2-BuOH = 10:1 yielded an even higher SFG signal. These results indicate that with the co-existence of O_2 in the gas phase, either the 2-BuOH molecules on Pt surface tended to be more ordered, or much higher surface 2-BuOH molecular density could be achieved.

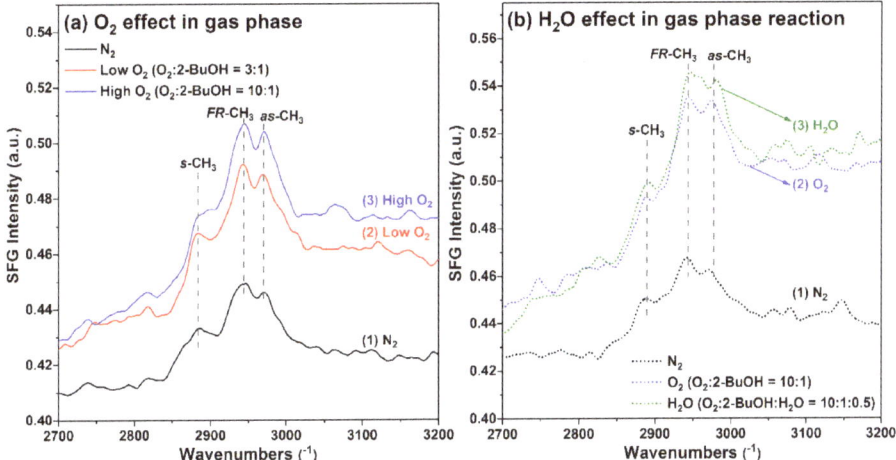

Figure 10. SFG spectra of gas phase 2-BuOH on Pt thin film at 80 °C (**a**) with varying ratios of oxygen to alcohol (i.e., N_2 only, low O_2 content with O_2:2-BuOH = 3:1, and high O_2 content with O_2:2-BuOH = 10:1), and (**b**) in the presence of water vapor (i.e., N_2 only, high O_2 content with O_2:2-BuOH = 10:1, and in the presence of H_2O with O_2:2-BuOH:H_2O = 10:1:0.5).

In order to provide insight into the influence of water vapor on 2-BuOH oxidation on Pt surface in the gas phase, the SFG spectra were also taken under reaction conditions with and without H_2O addition. As the results shown in Figure 10b indicate, again, the introduction of O_2 into system resulted in the sharp increase of SFG peak intensity, but the introduction of water vapor into system seemed to have little effect on the SFG peaks (even with some increased peak intensity for $-CH_{3,fr}$ and $-CH_{3,as}$). These results well support the catalytic data for the gas-phase 2-BuOH oxidation, in which the addition of water vapor had minimal effect on the reaction turnover rates. The SFG spectra analysis for 2-BuOH oxidation on Pt surface in the liquid phase is essential and highly recommended in future study to provide more information for alcohol oxidation chemistry in both phases at molecular level.

3. Materials and Methods

3.1. Pt Nanoparticle Synthesis and Encapulsation into Mesoporous Silica MCF-17

The Pt nanoparticles with sizes ranging from 2 to 9 nm were synthesized by a polyvinylpyrrolidone (PVP) assisted polyol process using ethylene glycol as reducing agent. The detailed procedures have been reported in our previous study [9,19–22]. Transmission electron microscopic (TEM) images showed that the as-synthesized Pt nanoparticles had narrow particle size distributions, and the high resolution TEM images with legible lattice fringes indicated that Pt(111) was the most favorable exposed surface.

Mesoporous silica MCF-17 with 20–30 nm pore size was used as support to immobilize the Pt nanoparticles. The as-synthesized Pt nanoparticles and MCF-17 support were mixed in ethanol solvent followed by sonication with 40 kHz, 80 W for 2 h. The MCF-17 supported Pt nanoparticles were collected by centrifugation followed by washing with ethanol for three times and drying at 80 °C overnight. Inductively coupled plasma atomic emission spectroscopy (ICP-AES) was used to determine the amount of Pt nanoparticles encapsulated in MCF-17. A known amount of Pt/MCF-17 catalysts were sonicated in ethanol first for good dispersion and then drop-casted on silica wafers followed by drying at room temperature for gas phase reactions. A known amount of Pt/MCF-17 catalysts were used, as it was for liquid phase reactions. The calculation of the active site number was based on the ICP-AES results and the size of the particles for turnover rates (TOF) calculation, assuming that Pt(111) surface was favorably exposed and every surface Pt atom was active in the catalytic alcohol oxidation reactions. The way of molecule adsorption on Pt surface, the influence of organic PVP capping agents and also the interface between Pt nanoparticle and mesoporous SiO_2 support was not taken into account for TOF calculation.

3.2. Catalytic Oxidation of Alcohols over Pt/MCF-17 Catalysts in Gas Phase and Liquid Phase

The gas phase alcohol oxidation reactions were performed in a gold-covered batch reactor, and the temperature of Pt/MCF-17 catalysts was precisely controlled by a boron nitride heater plate. Usually, in the typical gas phase reactions, 1.33 kPa of alcohols, 6.67 kPa of O_2 and 94.66 kPa of He were used, and the gas was circulated with a metal bellow pump. A gas chromatograph (GC) integrated with a flame ionization detector (FID) was used to monitor the gas composition online. All the alcohol conversions were kept under 10% during the data collection to calculate the turnover rates (TOF). For water effect study in the gas phase, 0.13–1.33 kPa of water vapor was introduced to the reaction chamber keeping the total pressure still as 102.66 kPa.

The liquid phase alcohol oxidation reactions were carried out in a Teflon-lined stainless-steel autoclave (Parr reactor, total volume of 100 mL). Usually, 15 mg catalysts were well dispersed in 15 mL liquid alcohols. The headspace of the reactor (85 mL) was purged and pressurized with O_2 (100 kPa for MeOH, EtOH, 1-PrOH, 2-PrOH, and 300 kPa for 2-BuOH) before the reaction started. The system was then heated to an elevated temperature and kept there for 3 h under continuous stirring. After the reaction, the mixture was cooled down, and the headspace gas was leaked to the GC-FID through an evacuated gas chamber for the gas phase product analysis. After the disassembly of the autoclave reactor, the catalyst was separated with centrifugation and the liquid phase was analyzed with GC-FID to obtain the liquid phase product information. For water effect study in the liquid phase, the ratio of H_2O in alcohols was adjusted to 0–75 V_{H2O}/V_{total} % while the total amount of liquid was kept at 15 mL. For low-concentration alcohol oxidation in the liquid phase, heptane was used as a diluter, and the ratio of alcohols to heptane was controlled at 1:1000.

3.3. SFG Spectral Analysis for 1-PrOH, 2-PrOH and 2-BuOH Oxidation in Gas and Liquid Phases

The SFG spectra taken for in situ catalytic 1-PrOH and 2-BuOH oxidations were carried out on Pt thin film prepared by electron beam deposition method [21]. The SFG spectra taken for in situ catalytic 2-PrOH oxidations were carried out on silica embedded 4 nm Pt nanoparticles deposited by deposition of Langmuir-Blodgett films onto sapphire prisms followed by calcination in air at 550 °C for 3 h [22]. All the SFG experiments were carried out in a system described in our previous work [32]. An active/passive mode-locked Nd:YAG laser (Continuum Leopard D10, 20 ps, 20 Hz, 1064 nm) was used to pump frequency conversion stages to generate two pulses, a visible pulse (532 nm, 130 µJ) and a tunable mid-IR pulse (2800 to 3600 cm^{-1}, 200 µJ). These pulses were overlapped at 62° (visible) and 45° (mid-IR) on the sample which had been deposited on the bottom of a quartz prism. Experiments were performed in the *ppp* polarization combination. The SFG signal was collected using a photomultiplier tube (PMT) accompanied by a gated integrator to improve the signal quality. The SFG spectra presented in this study had a resolution of 4 cm^{-1}, and each spectrum involved the average of at least 100 or 200 samplings to increase the signal-to-noise ratio for spectral analysis. Using a home-built cell, the surface was heated, and a recirculating mixture of the reaction gases/liquid was passed onto the Pt film/Pt nanoparticles. All the SFG spectra were normalized with the reference spectrum taken with a z-cut quartz.

3.4. DFT Calculation of Molecular Orientation of 1-PrOH and 2-PrOH on Pt Surface in Gas Phase and Liquid Phase

All geometry optimizations within the frame of DFT were performed with first-principles periodic system calculations using VASP software package [33]. The projector augmented wave (PAW) method was utilized to construct the basis set for the one-electron wave functions with plane-wave basis set limited by the cutoff energy of 400 eV [34]. For the constructed slabs, at least $3 \times 3 \times 1$ (for 1-PrOH) or $5 \times 5 \times 1$ (for 2-PrOH) Monkhorst-Pack k-point grid was used (depending on the geometrical dimensions). The Pt(111) slabs used were four layers thick and contained a minimum 1.7 nm (for 1-PrOH) or 1.2 nm (for 2-PrOH) vacuum space to exclude surface-surface interaction. The electronic steps were carried out with the energy convergence of 10^{-5} eV while the force convergence of ionic steps was set to be 5×10^{-2} eV/nm. For 1-PrOH, geometry optimization was performed using the PBE functional including Van der Waals interaction in DF approximation [35,36]. For 2-PrOH, geometry optimization was performed using the Perdew-Wang functional in generalized gradient approximation (GGA) [37]. The selected two constructed systems corresponded to two different concentrations of 1-PrOH and 2-PrOH molecules on Pt surface. At a low surface molecular coverage of 0.94 molecules/nm^2, the next-nearest distance between atoms of different molecules was more than 0.8 nm. At this separation distance, the alcohol molecules were considered to have minimal interaction at DFT level of theory. Therefore, we assumed that the corresponding molecular configuration represented the adsorption of alcohols from the low-pressure gas phase. On the other hand, the determined concentration of the 1-PrOH and 2-PrOH molecules in the liquid phase was about 3.94 molecules/nm^2, which was very close to the higher modeled surface concentration of 3.75 molecules/nm^2 that we could call liquid phase condition.

4. Conclusions

C1-C4 alcohol oxidation reactions including MeOH, EtOH, 1-PrOH, 2-PrOH and 2-BuOH in the gas phase and liquid phase over size-controlled Pt nanoparticles were systematically studied herein. The catalytic oxidation of alcohols in the gas phase showed much higher turnover rates than those in the liquid phase under the comparable surface density of reactants. Larger Pt nanoparticles exhibited higher turnover rates for both gas phase and liquid phase reactions, and in general, the liquid phase reaction showed very similar or higher selectivity to carbonyl compounds due to less complete oxidation reaction occurring forming CO_2 than the gas phase reaction. Interestingly, much lower apparent activation energy was observed for the alcohol oxidations in the liquid phase than those in the gas phase. In most cases, co-existing H_2O acted as a promoter for the liquid phase alcohol oxidations (especially for 1-PrOH and 2-PrOH) and as an inhibitor for the gas phase alcohol oxidations; however, totally the opposite H_2O effect was observed for 2-BuOH oxidation compared to other studied alcohols, which can be explained by hydrophilicity/hydrophobicity change for longer carbon chain alcohols. SFG spectra analysis suggested that significant change of molecular orientations was present for 1-PrOH and 2-PrOH on the Pt surface in the gas phase and liquid phase, and DFT calculation results confirmed that the alcohol molecules were mainly "lying down" on the Pt surface in the gas phase and "standing up" on Pt surface in the liquid phase. This led to totally distinct reaction kinetics and mechanisms for alcohol oxidations in different phases. SFG spectra analysis also revealed that the presence of O_2 could result in more ordered surface species or higher density of alcohol molecules on Pt surface in the gas phase 2-BuOH oxidation, and H_2O did not show an obvious influence on surface 2-BuOH molecules, explaining well why H_2O had little impact on turnover rates in the gas phase reaction. More systematic SFG spectra analysis and DFT calculation for the liquid phase 2-BuOH oxidation on Pt surface are necessary in future study.

Acknowledgments: This work was supported by the Director, Office of Basic Energy Sciences, Materials Science and Engineering Division of the U.S. Department of Energy under Contract No. DE-AC02-05CH11231.

Conflicts of Interest: The authors declare no conflict of interest.

References

1. DiCosimo, R.; Whitesides, G.M. Oxidation of 2-propanol to acetone by dioxygen on a platinized electrode under open-circuit conditions. *J. Phys. Chem.* **1989**, *93*, 768–775. [CrossRef]
2. Zhao, X.; Yin, M.; Ma, L.; Liang, L.; Liu, C.; Liao, J.; Lu, T.; Xing, W. Recent advances in catalysts for direct methanol fuel cells. *Energy Environ. Sci.* **2011**, *4*, 2736–2753. [CrossRef]
3. Besson, M.; Gallezot, P. Selective oxidation of alcohols and aldehydes on metal catalysts. *Catal. Today* **2000**, *57*, 127–141. [CrossRef]
4. Gauthier, E.; Benziger, J.B. Gas management and multiphase flow in direct alcohol fuel cells. *Electrochim. Acta* **2014**, *128*, 238–247. [CrossRef]
5. Gomes, J.F.; Bergamaski, K.; Pinto, M.F.S.; Miranda, P.B. Reaction intermediates of ethanol electro-oxidation on platinum investigated by sfg spectroscopy. *J. Catal.* **2013**, *302*, 67–82. [CrossRef]
6. Mallat, T.; Baiker, A. Oxidation of alcohols with molecular oxygen on solid catalysts. *Chem. Rev.* **2004**, *104*, 3037–3058. [CrossRef] [PubMed]
7. Jelemensky, L.; Kuster, B.F.M.; Marin, G.B. Multiple steady-states for the oxidation of aqueous ethanol with oxygen on a carbon supported platinum catalyst. *Catal. Lett.* **1994**, *30*, 269–277. [CrossRef]
8. Ciriminna, R.; Pandarus, V.; Béland, F.; Xu, Y.-J.; Pagliaro, M. Heterogeneously catalyzed alcohol oxidation for the fine chemical industry. *Org. Process Res. Dev.* **2015**, *19*, 1554–1558. [CrossRef]
9. Tatsumi, H.; Liu, F.; Han, H.-L.; Carl, L.M.; Sapi, A.; Somorjai, G.A. Alcohol oxidation at platinum–gas and platinum–liquid interfaces: The effect of platinum nanoparticle size, water coadsorption, and alcohol concentration. *J. Phys. Chem. C* **2017**, *121*, 7365–7371. [CrossRef]
10. Davis, S.E.; Ide, M.S.; Davis, R.J. Selective oxidation of alcohols and aldehydes over supported metal nanoparticles. *Green Chem.* **2013**, *15*, 17–45. [CrossRef]
11. Feng, J.; Ma, C.; Miedziak, P.J.; Edwards, J.K.; Brett, G.L.; Li, D.; Du, Y.; Morgan, D.J.; Hutchings, G.J. Au-pd nanoalloys supported on mg-al mixed metal oxides as a multifunctional catalyst for solvent-free oxidation of benzyl alcohol. *Dalton Trans.* **2013**, *42*, 14498–14508. [CrossRef] [PubMed]
12. Papes Filho, A.C.; Maciel Filho, R. Hybrid training approach for artificial neural networks using genetic algorithms for rate of reaction estimation: Application to industrial methanol oxidation to formaldehyde on silver catalyst. *Chem. Eng. J.* **2010**, *157*, 501–508. [CrossRef]
13. Slot Thierry, K.; Eisenberg, D.; van Noordenne, D.; Jungbacker, P.; Rothenberg, G. Cooperative catalysis for selective alcohol oxidation with molecular oxygen. *Chem. Eur. J.* **2016**, *22*, 12307–12311. [CrossRef] [PubMed]
14. Vinod, C.P.; Wilson, K.; Lee Adam, F. Recent advances in the heterogeneously catalysed aerobic selective oxidation of alcohols. *J. Chem. Technol. Biotechnol.* **2011**, *86*, 161–171. [CrossRef]
15. Kopylovich, M.N.; Ribeiro, A.P.C.; Alegria, E.C.B.A.; Martins, N.M.R.; Martins, L.M.D.R.S.; Pombeiro, A.J.L. Chapter Three-Catalytic oxidation of alcohols: Recent advances. In *Advances in Organometallic Chemistry*; Pérez, P.J., Ed.; Academic Press: Cambridge, MA, USA, 2015; Volume 63, pp. 91–174.
16. Enache, D.I.; Edwards, J.K.; Landon, P.; Solsona-Espriu, B.; Carley, A.F.; Herzing, A.A.; Watanabe, M.; Kiely, C.J.; Knight, D.W.; Hutchings, G.J. Solvent-free oxidation of primary alcohols to aldehydes using Au-Pd/TiO$_2$ catalysts. *Science* **2006**, *311*, 362–365. [CrossRef] [PubMed]
17. Sharma, A.S.; Kaur, H.; Shah, D. Selective oxidation of alcohols by supported gold nanoparticles: Recent advances. *RSC Adv.* **2016**, *6*, 28688–28727. [CrossRef]
18. Mallat, T.; Baiker, A. Oxidation of alcohols with molecular oxygen on platinum metal catalysts in aqueous solutions. *Catal. Today* **1994**, *19*, 247–283. [CrossRef]
19. Wang, H.; An, K.; Sapi, A.; Liu, F.; Somorjai, G.A. Effects of nanoparticle size and metal/support interactions in pt-catalyzed methanol oxidation reactions in gas and liquid phases. *Catal. Lett.* **2014**, *144*, 1930–1938. [CrossRef]
20. Sapi, A.; Liu, F.; Cai, X.; Thompson, C.M.; Wang, H.; An, K.; Krier, J.M.; Somorjai, G.A. Comparing the catalytic oxidation of ethanol at the solid–gas and solid–liquid interfaces over size-controlled pt nanoparticles: Striking differences in kinetics and mechanism. *Nano Lett.* **2014**, *14*, 6727–6730. [CrossRef] [PubMed]

21. Liu, F.; Han, H.-L.; Carl, L.M.; Zherebetskyy, D.; An, K.; Wang, L.-W.; Somorjai, G.A. Catalytic 1-propanol oxidation on size-controlled platinum nanoparticles at solid-gas and solid-liquid interfaces: Significant differences in kinetics and mechanisms. *J. Phys. Chem. C* **2018**. [CrossRef]
22. Wang, H.; Sapi, A.; Thompson, C.M.; Liu, F.; Zherebetskyy, D.; Krier, J.M.; Carl, L.M.; Cai, X.; Wang, L.-W.; Somorjai, G.A. Dramatically different kinetics and mechanism at solid/liquid and solid/gas interfaces for catalytic isopropanol oxidation over size-controlled platinum nanoparticles. *J. Am. Chem. Soc.* **2014**, *136*, 10515–10520. [CrossRef] [PubMed]
23. Saturated Vapor Pressure. Available online: http://ddbonline.ddbst.com/AntoineCalculation/AntoineCalculationCGI.exe (accessed on 1 February 2018).
24. NIST Chemistry WebBook. Available online: https://webbook.nist.gov/cgi/inchi?ID=C71238&Mask=4&Type=ANTOINE&Plot=on (accessed on 1 February 2018).
25. Kemme, H.R.; Kreps, S.I. Vapor pressure of primary n-alkyl chlorides and alcohols. *J. Chem. Eng. Data* **1969**, *14*, 98–102. [CrossRef]
26. Kobayashi, H.; Higashimoto, S. DFT study on the reaction mechanisms behind the catalytic oxidation of benzyl alcohol into benzaldehyde by O_2 over anatase TiO_2 surfaces with hydroxyl groups: Role of visible-light irradiation. *Appl. Catal. B Environ.* **2015**, *170–171*, 135–143. [CrossRef]
27. Welty, J.R.; Wicks, C.E.; Wilson, R.E.; Rorrer, G.L. *Fundamentals of Momentum, Heat, and Mass Transfer 5th Edition*; John Wiley & Sons, Inc.: New York, NY, USA, 2007; ISBN 2900470128687.
28. Ferrell, R.T.; Himmelblau, D.M. Diffusion coefficients of nitrogen and oxygen in water. *J. Chem. Eng. Data* **1967**, *12*, 111–115. [CrossRef]
29. Alger, D.B. The water solubility of 2-butanol: A widespread error. *J. Chem. Educ.* **1991**, *68*, 939. [CrossRef]
30. Mullen, G.M.; Zhang, L.; Evans, E.J.; Yan, T.; Henkelman, G.; Mullins, C.B. Oxygen and hydroxyl species induce multiple reaction pathways for the partial oxidation of allyl alcohol on gold. *J. Am. Chem. Soc.* **2014**, *136*, 6489–6498. [CrossRef] [PubMed]
31. Mullen, G.M.; Zhang, L.; Evans, E.J.; Yan, T.; Henkelman, G.; Mullins, C.B. Control of selectivity in allylic alcohol oxidation on gold surfaces: The role of oxygen adatoms and hydroxyl species. *Phys. Chem. Chem. Phys.* **2015**, *17*, 4730–4738. [CrossRef] [PubMed]
32. Thompson, C.M.; Carl, L.M.; Somorjai, G.A. Sum frequency generation study of the interfacial layer in liquid-phase heterogeneously catalyzed oxidation of 2-propanol on platinum: Effect of the concentrations of water and 2-propanol at the interface. *J. Phys. Chem. C* **2013**, *117*, 26077–26083. [CrossRef]
33. Kresse, G.; Furthmüller, J. Efficient iterative schemes for ab initio total-energy calculations using a plane-wave basis set. *Phys. Rev. B* **1996**, *54*, 11169–11186. [CrossRef]
34. Kresse, G.; Joubert, D. From ultrasoft pseudopotentials to the projector augmented-wave method. *Phys. Rev. B* **1999**, *59*, 1758–1775. [CrossRef]
35. Klimeš, J.; Bowler, D.R.; Michaelides, A. Van der waals density functionals applied to solids. *Phys. Rev. B* **2011**, *83*, 195131. [CrossRef]
36. Román-Pérez, G.; Soler, J.M. Efficient implementation of a van der waals density functional: Application to double-wall carbon nanotubes. *Phys. Rev. Lett.* **2009**, *103*, 096102. [CrossRef] [PubMed]
37. Perdew, J.P.; Chevary, J.A.; Vosko, S.H.; Jackson, K.A.; Pederson, M.R.; Singh, D.J.; Fiolhais, C. Atoms, molecules, solids, and surfaces: Applications of the generalized gradient approximation for exchange and correlation. *Phys. Rev. B* **1992**, *46*, 6671–6687. [CrossRef]

© 2018 by the authors. Licensee MDPI, Basel, Switzerland. This article is an open access article distributed under the terms and conditions of the Creative Commons Attribution (CC BY) license (http://creativecommons.org/licenses/by/4.0/).

MDPI
St. Alban-Anlage 66
4052 Basel
Switzerland
Tel. +41 61 683 77 34
Fax +41 61 302 89 18
www.mdpi.com

Catalysts Editorial Office
E-mail: catalysts@mdpi.com
www.mdpi.com/journal/catalysts

www.ingramcontent.com/pod-product-compliance
Lightning Source LLC
LaVergne TN
LVHW071957080526
838202LV00064B/6776